Springer Geography

For further volumes:
http://www.springer.com/series/10180

Stefan Mann
Editor

The Future of Mountain Agriculture

Springer

Editor
Stefan Mann
Socioeconomics
Agroscope
Ettenhausen
Switzerland

ISBN 978-3-642-33583-9 ISBN 978-3-642-33584-6 (eBook)
DOI 10.1007/978-3-642-33584-6
Springer Heidelberg New York Dordrecht London

Library of Congress Control Number: 2012952138

Printed on acid-free paper

Springer is part of Springer Science+Business Media (www.springer.com)

Contents

Introduction

Stefan Mann

I recently told a friend of mine about my effort to edit a book about mountain agriculture. "Oh is it about management issues?", was the question of the professor of institutional economics, and from her perspective this was a very reasonable guess: Starting with a study about Alpine Pasture management (Ostrom 1990) which contributed to Eleanor Ostrom's award of the Nobel Prize in 2009, mountain agriculture has recently been a very fertile test bed for different institutional solutions of property management (Pandit and Thapa 2004; Turkelboom et al. 2004). In remote valleys, there often is a tradition of shared property or at least shared utilization of certain resources, and their thorough analysis allows to draw broader conclusions about the feasibility of different governance mechanisms.

However, very many other discourses also use mountain agriculture as a case in point with a fortune of empirical material. Landscape economists, for example, often work on mountainous landscape as their cases to prove the existence of positive externalities of conservation (Cobbing and Slee 1993; Campbell 2007). Similarly, ecologists who are interested in the impacts of climate change on biodiversity and agriculture often concentrate on mountain regions for their observations and forecasts (Beniston 2003; Huber et al. 2005; Giupponi et al. 2006). These three discourses (and some others) are held intensively and with frequent references to some mountain regions, but with little interaction in between them.

There is a similar situation with discourses that focus on a particular region, although their contents may be broader. The future of agriculture in the Alps, for example, is a often discussed subject (Flury et al. 2004; Mack et al. 2008), in

S. Mann (✉)
Socioeconomics Research Group, Swiss Federal Research Station Agroscope,
Tänikon 1, 8356 Ettenhausen, Switzerland
e-mail: stefan.mann@art.admin.ch

S. Mann (ed.), *The Future of Mountain Agriculture*, Springer Geography,
DOI: 10.1007/978-3-642-33584-6_1,

which Austrian, Swiss, Italian and other scientists are reasonably well interlinked. However, there is hardly an exchange between them and, say, the discourse on agriculture in the Rocky Mountains or even the Himalaya.

Therefore, the reply to my friend was: "No, it is far less methodological. I am just interested whether the system of mountain farming is going to persist. The land in mountainous areas is often steep, the soils are partly poor, and weather conditions not always favorable to agricultural production. In a globalised economy, it is worthwhile to ask whether the system of mountain farming, in its stunning diversity, will in general continue to provide food, labor and ecosystem services."

And this is what this book is about. There are many things which we know about the future of agriculture. We know that, a few decades from now on, nine or ten billion people need to be fed. And we know that fertile lowlands which we find from California in the North-West to New Zealand in the South-East will play a crucial role to fulfill this requirement. But will we also need the mountainous regions on the major continents for the provision of milk, meet and possibly even arable products? Or can we imagine mountain landscapes in which agricultural practices are only preserved in a few museums?

The question can also be formulated from an environmental perspective: We know that the biodiversity hotspots of the world include many mountainous regions such as the Tropical Andes and the Caucasus (Myers et al. 2000). While it is unlikely that the efforts to preserve this wealth will decrease, it is open in which way biodiversity will be preserved. Will rangers guard protected areas in which all attention is focused on the provision of ecosystem services? Or will strategies be pursued where food production on a sustainable level and biodiversity protection will be linked through appropriate agricultural systems?

It is exciting to read the single contributions about the single mountain regions in this book. But it is even more revealing to compare them in order to discover both parallels and differences. Doing so will question a lot of stereotypes and prejudices. From my European perspective, for example, it was clear that mountain agriculture, at least in the developed world, would undergo structural change where many farms would be given up. The only question would be whether the land of affected farms would also be abandoned or would be transferred to the management of other farms. While such processes are very well described by *Feliu Lopez-i-Gelats* in his chapter about the Pyrenees, *Sarah Cline* describes almost a reverse process for the Rocky Mountains: The number of farms there is actually increasing. Many of the current ranches in the area are being subdivided and used for smaller operations.

However, the common ground of the North American and the European perspective is that multifunctionality is coming on the stage, as clearly is described in the chapter about the Alps by *Christian Flury, Robert Huber* and *Erich Tasser*. Multifunctionality can be described as a codeword for those who emphasize the many contributions of the farm sector outside plain food production. In the Rocky Mountains, the Pyrenees and the Alps, food production still occurs, but is increasingly sidelined by the demands on land functions regarding recreation, landscape and biodiversity.

The link between conserving nature and producing food is not restricted to developed countries, as *Jeffrey Alwang, George Norton, Victor Barrera* and *Ruben Botello* show in their contribution. The difference is, however, that hardly anybody appears to question the core role of food production of the agricultural systems in the Andes, the Kilimanjaro, the Atlas Mountains or the Himalaya. Both Switzerland and Tajikistan are mountainous countries that strongly depend on food imports. For the Swiss, this may be annoying because not all imports fulfill the high quality standards to which Swiss consumers are adapted. But Tajiks have been severely threatened in their daily subsistence after the worldwide food price hike occurred, and their top priority is to foster agricultural production on the area suitable for plant growth and grazing, as *Kamilijon Akramov* describes in his chapter.

There is more than one bottleneck hampering the productivity of mountain agriculture. *Werner Doppler* and *Krishna Bahadur* describe how infrastructure can be a crucial point. Improved roads, for example, will enable many farming families in peripheral regions to switch to more productive farming systems. Another important factor is the dissemination of know-how. *Jon Hellin* shows how networks of farming consultants (or consulting farmers) contribute to spread innovative methods of production, processing and marketing.

Unfortunately, mountain agriculture is not only challenged by opportunities, but also by threats. A growing population and altering social structures change institutional regimes that have worked over centuries, as *Bernadette Montanari* describes for the thyme harvest. And last but not least, global warming is a considerable threat for the viability of mountain agriculture as we know it today, as highlighted by *Francis Mulangu* and *David Kraybill*. Production technologies, including the use of irrigation, will be seriously affected by an altered climate and needs a well-planned process of adaptation.

Having worked myself through all the fascinating chapters of this book, I feel that the question that drove me in editing this book may have been naïve and shaped strongly from my European perspective. Of course mountain agriculture will persist. While rich countries will use the opportunity to increasingly link food production with other amenities that can be enjoyed in mountainous regions, there are few alternatives to continued food production for the large peasant population in Asia, Africa and South America. Climate change and an increased population pressure is not going to make this an easy task. But the growing knowledge about the important role of sustainable production methods, of innovation adoption, of marketing adapted to consumer demands and of a suitable political framework enable academics and practitioners to adapt and improve agricultural systems in the different region to the needs of future generations.

References

Beniston M (2003) Climatic change in mountain regions: a review of possible impacts. Clim Change 59(1–2):5–31

Campbell D (2007) Willingness to pay for rural landscape improvements: combining mixed logit and random-effects models. J Agric Econ 58(3):467–483

Cobbing P, Slee B (1993) A contingent valuation of the Mar Lodge Estate, Cairngorm Mountains, Scotland. J Environ Planning Manage 36(1):65–72

Flury C, Gotsch N, Rieder P (2004) Zukunft im Wandel: Erwartete Entwicklung der Landwirtschaft im Alpenraum. Agrarwirtschaft und Agrarsoziologie 1(2004):55–74

Giupponi C, Ramanzin M, Sturaro E, Fuser S (2006) Climate and land use changes, biodiversity and agri-environmental measures in the Belluno province, Italy. Environ Sci Policy 9(2):163–173

Huber UM, Bugmann HKM, Reasoner MA (2005) Global change and mountain regions—an overview of current knowledge. Springer, Heidelberg

Mack G, Walter T, Flury C (2008) Entwicklung der Alpung in der Schweiz: Ökonomische Bedeutung und ökologische Auswirkungen. Yearb Socioecon Agric 2008:259–300

Myers N, Mittermeier RA, Mittermeier CG, da Fonseca GAB, Kent J (2000) Biodiversity hotspots for conservation priorities. Nature 403:852–858

Ostrom E (1990) Governing the commons. Cambridge University Press, Cambridge

Pandit BH, Thapa GB (2004) Poverty and Resource Degradation Under Different Common Forest Resource Management Systems in the Mountains of Nepal. Society & Natural Resources 17(1):1–16

Turkelboom F, Ghafouri M, Sarreshtehdari A (2004) Integrated watershed management for the dry mountains of Iran—adapting worldwide lessons learnt to Iranian conditions. In: Farahani H, Oweis T, Siadat H, Abbasi F, Bruggeman A, Anthofer J, Turkelboom F (eds) Improving on-farm agricultural water productivity in the Karkheh river basin. ICARDA, Aleppo

Land Use and Landscape Change in the Rockies: Implications for Mountain Agriculture

Sarah A. Cline

1 Introduction

Agriculture has historically been an important economic driver in many states in the Rocky Mountain region of the United States.[1] Over half of all grassland pasture and range in the United States is located in the Mountain Region, and grassland pasture and range makes up over half of all land in the Mountain Region (Nickerson et al. 2011). While the mix of agricultural production varies throughout the region, cattle production is one of the top commodities in terms of value in many Rocky Mountain States. Globally, the outlook for livestock products is promising, with recent FAO projections indicating that beef prices are expected to increase 18 % by 2020, compared to current levels (OECD-FAO 2011). However, even with an expectation of increasing beef prices, the area available for livestock grazing in the United States has been declining over time. The percentage of total land area used for grazing decreased by 24 % from 1949 to 2007 (Nickerson et al. 2011). The Mountain Region has contributed to this decline, with a decrease of around 11 % over the period.

Sarah Cline is an economist with the U.S. Department of the Interior, Office of Policy Analysis. The views expressed here are those of the author and not of the Department of the Interior or the United States Government.

[1] Throughout this paper, we focus on the U.S. portion of the Rocky Mountain range, and use the U.S. Bureau of Census definition of the Rocky Mountain region, which includes the following states: Idaho, Montana, Wyoming, Nevada, Utah, Colorado, Arizona, and New Mexico. References to the Mountain Region from USDA sources include the same list of states.

S. A. Cline (✉)
U.S. Department of the Interior, Office of Policy Analysis, Washington, DC, USA
e-mail: clinesarah@gmail.com

S. Mann (ed.), *The Future of Mountain Agriculture*, Springer Geography,
DOI: 10.1007/978-3-642-33584-6_2,

Although agriculture varies throughout the Rocky Mountain States, many mountain farmers face similar challenges. Much of the previous literature has noted that ranchlands in the west do not offer competitive profit or return on investment; some yield negative returns. It is argued that many ranches across the west are marketed or sold at above their agricultural value (Gosnell and Travis 2005). The high prevalence of publically-owned lands throughout much of the region contributes to the scarcity of land that is available for purchase and development. About 40 % of all federal lands in the United States are located in the Rocky Mountain region (Nickerson et al. 2011). In addition, the abundance of natural amenities including scenic mountain vistas, recreational opportunities, and clean air and water, have made many of these communities desirable for migrants, retirees, and tourism. The resulting development pressures and high land prices have made it more difficult to maintain viable agricultural operations throughout the region.

However, even with this increased demand for residential and commercial land uses, significant public benefits are derived from the natural amenities provided by agricultural landscapes in many areas of the Rocky Mountain States. The maintenance of working ranches and farms can help to avoid the increasingly fragmented landscapes in mountain communities, providing economic and ecological benefits including wildlife habitat and natural resource provision. Previous research has shown that the preservation of agricultural working landscapes and associated way of life is valuable to residents and tourists in many mountain communities in the region (Rosenberger and Walsh 1997; Rosenberger and Loomis 1999; Orens and Seidl 2009; Cline and Seidl 2009; Ellingson et al. 2011; Magnan et al. 2012).

In order to capitalize on these values placed on working landscapes, mountain farmers have begun to consider various options to enhance the viability of farms and ranches. One option is the development of agritourism operations to help supplement revenues. Other policy options such as conservation easements and subsidies can provide additional opportunities for many mountain farmers to maintain viable ranching and farming operations in the face of pending development.

This chapter discusses the challenges facing mountain farmers in the Rockies, focusing on the viability of ranching in mountain communities facing development pressure from residential and commercial land uses. The next section outlines historical trends in agricultural production in the region. The following section discusses trends in land use and land use change in the Rocky Mountains. A summary of the literature measuring the non-market benefits supplied by ranchlands in the Rocky Mountains is provided, followed by a discussion of potential ways to capitalize on these values and possible policy options to improve the future viability of ranches throughout the Rocky Mountain region. The final section provides conclusions and policy implications.

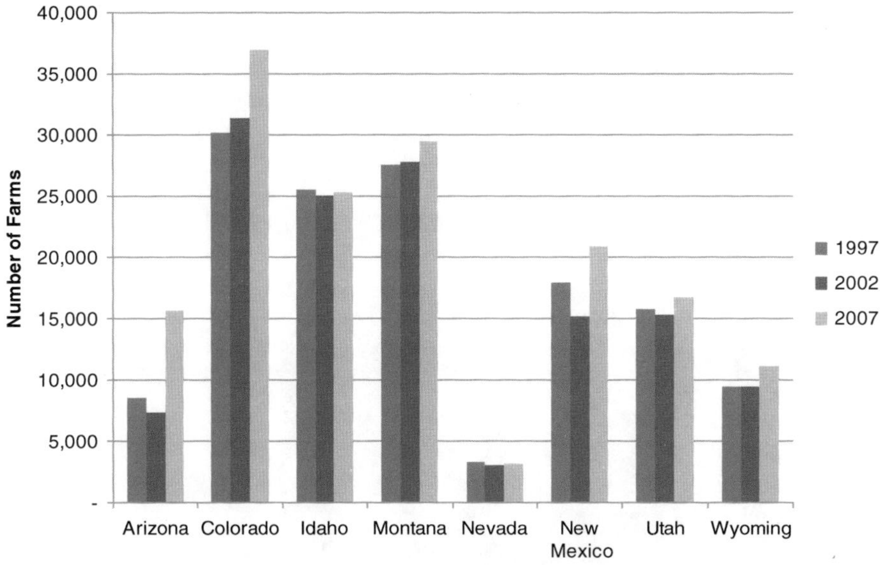

Fig. 1 Number of Farms in Rocky Mountain States, 1997–2007

2 Agricultural Production Trends in the Rocky Mountain Region

The Rocky Mountains are a major mountain range in North America, stretching more than 3,000 miles (4,830 km) from British Columbia in western Canada, to New Mexico in the southwestern United States. The highest point in the Rockies is Mount Elbert in Colorado, at 14,440 ft (4,401 m). A large part of land in the region is protected by state and national forests and parks. Economic resources in the region are diverse, including timber, minerals, agriculture, and tourism.

Agriculture has historically been an important economic driver in parts of the Rocky Mountain region. Trends in U.S. Agricultural Census data in recent years show an increase in the total number of farms in many of the Rocky Mountain States (Fig. 1), while the average size of farms decreased in many states between 1997 and 2007 (Fig. 2). The number of farms remained relatively stable, with slight decreases in some states, between 1997 and 2002. Between 2002 and 2007, the number of farms increased throughout the region, with the largest percentage increases in Arizona, New Mexico, Colorado, and Wyoming. The average farm size increased slightly in all states but Colorado between 1997 and 2002. In recent years, average farm size has decreased more significantly in many states, with a percentage decrease of over 10 % in Arizona, New Mexico, Wyoming, Colorado, Utah, and Nevada between 2002 and 2007.

The market value of livestock and poultry production was greater than crop production in all states in the Rocky Mountain region except Arizona in 2007

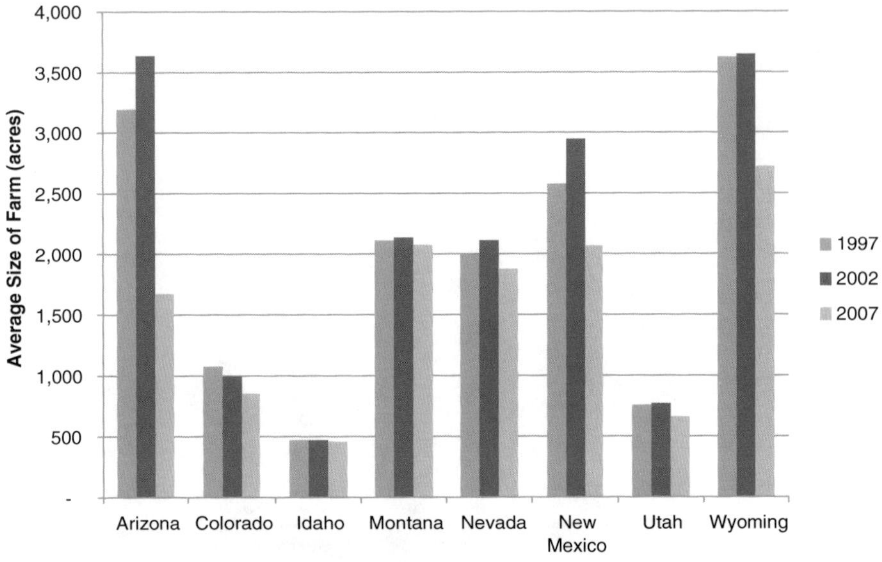

Fig. 2 Average Size of Farm in Rocky Mountain States, 1997–2007

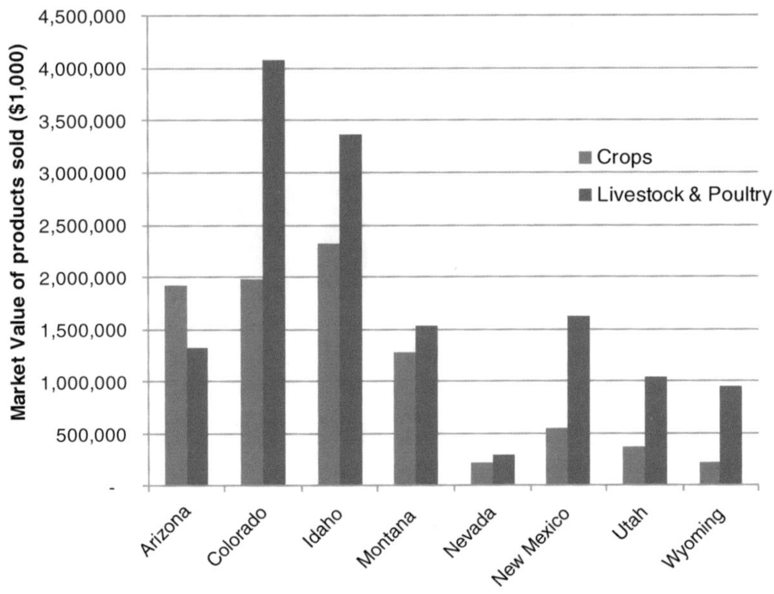

Fig. 3 Market Value of Products Sold in Rocky Mountain States, 2007

(Fig. 3). Market value was largest in Colorado, with livestock and poultry almost twice as large as crop production. The value of livestock and poultry production was also significantly greater than crop production in New Mexico, Utah, and Wyoming.

Recent statistics show that the majority of agricultural land in many states in the region is used for pasture. While the mix of agricultural production varies throughout the region, cattle production is one of the top commodities in terms of value in many Rocky Mountain States. Particularly within the mountain areas of these states, grazing and ranching have traditionally made up a large part of agricultural activity.

As shown in the state level statistics above, on average, farms in the region are getting smaller, while the total number of farms is increasing. The reduction in size of farming operations is occurring in the mountain areas of these states as well. Competition for land from other uses has led to increased fragmentation of the agricultural landscape. The next section discusses these trends in more detail.

3 Land Use and Landscape Change

A significant portion of all land in the United States is used for agricultural purposes. The most recent data available from the United States Department of Agriculture (USDA) indicates that in 2007, grassland and range made up 27 % of total land area, and cropland made up another 18 %, while urban uses made up only 3 %. Land use for urban and rural residential uses has been increasing significantly, however, with a quadrupling of urban land uses between 1945 and 2007 throughout the United States (Nickerson et al. 2011). In recent decades, growth in urban land use has continued to increase, with a 17 % increase since 1990, and a 2 % increase between 2002 and 2007 (Nickerson et al. 2011).

Residential land uses have been increasing in many rural areas as well. Rural residential acreage outside urban areas increased 29 % between 1997 and 2002, and 10 % between 2002 and 2007 (Nickerson et al. 2011). This increasing development in the rural–urban fringe has threatened agricultural landscapes in many rural communities throughout the United States. Between 1992 and 2001, an average of 2.2 million acres of farmland in the United States was converted to urban uses per year (Nickerson and Hellerstein 2007).

Similar trends have been ongoing in the Rocky Mountain region in recent years. Land acreage in urban uses increased by 2 % for the Rocky Mountain States between 2002 and 2007. Population growth in the Rocky Mountain States has contributed to regional land use changes. The rapid population growth occurring in the Rocky Mountain West and its implications for agricultural production have been discussed for decades (Diemer 1979).

These trends have affected land use change in mountain areas in the Rocky Mountain States. Population growth and an increased number of migrants moving to mountain areas, along with greater levels of tourism to the area have led to increased demand for the limited private lands in many counties. In many cases, large land areas used for agriculture or ranching have been split into smaller parcels and sold for hobby farms, "ranchettes", or other types of development. This type of development can lead to a more fragmented landscape that can affect the local community. Riebsame et al. (1996) noted the landscape change during the 1990s throughout the Colorado mountains. Demand from residential and commercial interests in the area has led to land use conversion, primarily from agricultural to residential uses, due in large part to demand for second homes. A number of other studies have documented similar landscape changes throughout the Rocky Mountains.

Gosnell and Travis (2005) conducted three case studies of ranch sales in mountain counties between 1990 and 2000. The three locations included Routt County, CO, a resort area; Sublette County, WY which is near the resort area of Jackson Hole, WY; and Carbon County, MT which is a more rural location. The trends in ranch sales varied in the three counties over the time period of the study. Routt County showed the largest number of sales and the largest acreage sold of the three case studies. Sublette County saw the largest percentage of ranches change hands during the study period. The largest percentage of sales and total acreage sold (over half) went to amenity buyers in both Routt and Sublette Counties. Only 15 % of ranches in Carbon County changed hands during the study period, with a smaller percentage of total sales going to amenity buyers. The results of these case studies suggest that amenity buyers have a significant effect in many areas of the Rocky Mountain West, although there may be a tipping point of number of sales or acreage sold before a large-scale landscape change occurs. The location of the county with respect to resorts or other amenity buyers may influence the speed with which landscape change occurs.

Gosnell et al. (2006) conducted another analysis of ranchland ownership change from 1990 to 2000 in the area around Yellowstone National Park, encompassing 18 million acres in Montana, Idaho, and Wyoming. They found that most entrants into the land market were not traditional farmers and ranchers, leading to a transition from ranchers that were mainly involved in meat production to a more diverse set of landowners including: absentee owners focused on amenity or conservation values instead of or in addition to livestock, investors, and land developers. Most of the land that changed hands in the counties being studied actually stayed in large, undeveloped parcels. About 22 % of the large agricultural landholdings changed hands during the period, with sales dominated by amenity buyers, traditional ranchers, and investors, in that order. Developers and conservation organizations made up much smaller percentages of the total sales. The results of this study suggest that fragmentation due to exurban development is proceeding more slowly in more rural areas than in areas with resorts and near urban areas. Although some parts of the area already have as many (or more) non-traditional ranchers as traditional ones. In order to help reduce fragmentation, prime targets for conservation easements in the future could be in areas where there is a fragmented landscape and greater demand for development.

Overall, these studies show that in many areas of the Rocky Mountain region, demand for land on the valley floor by non-traditional ranchers is beginning to lead to farms and ranches being divided into smaller parcels. The rate of conversion and number of farms affected may depend on a number of factors, including the location of the site. More rural locations, and those further away from resort or other high amenity locations seem to be at less risk for development currently than those areas closer to resorts or urban areas.

4 Natural Amenities and Valuation of Agricultural Landscapes

As discussed above, agricultural lands in many areas of the Rocky Mountains have faced increasing competition from residential and commercial land uses in the recent decades. When comparing the value of different land uses, the productive values of each use are often considered. However, valuing agricultural lands at the value of their livestock or crop production is likely to understate the total economic benefit these lands provide to the local community. Agricultural working landscapes provide many ecosystem and cultural benefits to residents and visitors alike in mountain communities. Specific benefits can include habitat for wildlife and endangered species, recreational opportunities, water quality and quantity regulation, and open space, as well as cultural benefits related to ranching and rural lifestyles. Coupal et al. (2004) discuss the importance of private and public ranchlands in supporting habitat for wildlife in Wyoming, and point out that although ranchers can receive some benefits such as access fees, most economic impacts accrue to other businesses related to tourism. Although the values that individuals place on the environmental and cultural characteristics of working landscapes are not included in the market price of agricultural lands, understanding the value of these non-market benefits is important for policy decisions about land use planning in mountain communities.

The total economic value of ranching includes not only the market benefits of the goods provided, but also the non-market benefits of the ecosystem and cultural goods and services discussed above. Seidl (2006) provides an overview of the different categories of non-market economic value with respect to ranchlands in mountain communities. Use values, generally described as the values that individuals receive from being in direct contact with the resource, can be categorized as consumptive use or non-consumptive use. Activities such as hiking, fishing, and wildlife viewing undertaken on agricultural lands can provide consumptive use values, while non-consumptive use values can be derived from viewing pastoral landscapes, water filtration, and flood control. Non-use values, or values that individuals place on a good even though they may not have direct contact with it, are generally categorized as existence, bequest or option values. Examples of existence values (benefits that an individual derives from a good that they will not directly experience) obtained from ranchlands could include habitat for threatened or endangered species. Bequest values are derived from providing the benefits for future generations; and in the case of agricultural lands may include maintenance of rural lifestyles, and well as the provision of habitat for wildlife and endangered species. Option values are derived from maintaining the possibility of using the resource in the future, which in the case of agricultural lands could include future land uses that may be precluded if lands were converted to residential or commercial uses.

Non-market valuation methods are often used to estimate values for environmental goods and services such as those provided by agricultural lands. Revealed preference methods are able to uncover or "reveal" the value that individuals place on an environmental good by observing their behavior in a related market. For example, travel cost (TC) analysis is often used to estimate values for different site characteristics by observing (generally through survey data) the amount individuals are willing to incur in travel costs to visit sites with different attributes or different levels of environmental quality. Hedonic analysis is another revealed preference technique that isolates the value of an environmental good out of a bundle of characteristics of a particular good, such as a house. Stated preference methods take a different approach by asking individuals what value they place on a particular environmental good or service. Contingent valuation (CV) uses survey methodology to derive values for different levels of environmental quality by asking respondents how much they would be willing to pay for the provision of or the change in the quality or quantity of a given environmental good. The contingent behavior (CB) method is another stated preference technique that is used to assess how individuals would change their behavior given a change in price or environmental quality. Individuals are asked to state how their behavior (for example trips taken per year or season) might change given a particular change in environmental quality, price or access to the site.

With the conversion of agricultural lands to other uses, a number of research studies have been conducted to attempt to estimate the value of the non-market attributes provided by these lands (see Bergstrom and Ready 2009 for an overview). Most of these studies have focused on measuring the value of the open space amenities that these lands provide. Several authors have used hedonic methods to estimate the value of agricultural or ranchland open space to residents (Geoghegan 2002; Ready et al. 1997), while others have used hedonic methods to estimate these values for tourists by looking at rental prices for cottages in rural areas (Vanslembrouck et al. 2005; Le Goffe 2000). A number of other authors have estimated the value of agricultural lands to residents using contingent valuation methods (Bergstrom et al. 1985; Bowker and Didychuk 1994; Ready et al. 1997; Rosenberger and Walsh 1997). Other studies have been conducted to estimate the value of farm and ranchlands to tourists using contingent behavior and travel cost information (Rosenberger and Loomis 1999; Orens and Seidl 2009).

These methods have been used in several studies to estimate the non-market benefits of agricultural lands in the Rocky Mountain region. A number of studies were conducted in mountain communities in Colorado beginning in the 1990s to estimate the value of open space provided by ranchlands to community residents and tourists (Rosenberger and Walsh 1997; Rosenberger and Loomis 1999; Orens and Seidl 2009; Cline and Seidl 2009; Ellingson et al. 2011; Magnan et al. 2012).

Table 1 Non-market valuation studies of agricultural land in the Rocky Mountain region

Citation	Location	Method[a]	Population	Value[b]	Years
Magnan et al. (2012)	Routt County, CO	CV	Resident	$220	2004
Orens and Seidl (2009)	Gunnison County, CO	CB	Tourist	$0.002	2003
Rosenberger and Loomis (1999)	Routt County, CO	CB	Tourist	$0	1993
Rosenberger and Walsh (1997)	Routt County, CO	CV	Resident	$107–256	1993
Ellingson et al. (2011)	Routt County, CO	CV/CB	Tourist	$129	2005
Cline and Seidl (2009)	Chaffee County, CO	CB	Tourist	$5–73	2007

[a] *CV* Contingent Valuation, *CB* Contingent Behavior

[b] Values in this table are not directly comparable due to different methodologies and different units across studies

Table 1 provides a summary of non-market valuation studies estimating values for amenities associated with agricultural lands in the Rocky Mountain region. The table shows estimated willingness to pay to protect open space and other details for each study, although values across the studies should not be directly compared due to different methodologies and different measurement units across studies.

Studies were conducted in the 1990s and 2000s to estimate the value of ranchland open space to residents of Routt County, Colorado. Routt County is home to the town of Steamboat Springs and a well-known ski resort. Routt County lost around 20 % of its valley ranchland between 1990 and 1995, leading to interest from various stakeholder groups about options for maintaining agricultural lands in the area. This interest led to a study by Rosenberger and Walsh (1997) which estimated the value of ranchland open space to county residents using the contingent valuation method. Although their results showed that local residents had a positive willingness to pay for the non-market benefits of open space in the region (with a range of $107–256 annually per household depending upon the percentage of acres protected), the regional marginal values were not sufficient to dominate land prices for development uses. A decade later, Magnan et al. (2012) used information from the 1993 survey and a survey from 2004 to determine if changes in population and demographics affected the residents' willingness to pay. The study used the CV method to determine residents' WTP to protect ranch open space. The mean WTP was found to be $220 in 2004. The study found that concern for open space and WTP to protect open space did not change significantly over the 10 year period. This empirical result contradicts the assertion that urban to rural migration will negatively affect support for agricultural open space.

Several studies have also been conducted that estimate the WTP of tourists for agricultural open space in Colorado mountain communities. Rosenberger and Loomis (1999) assessed the value of ranchland open space to summer tourists in Routt County, Colorado. This area includes the resort area of Steamboat Springs, which draws a large number of tourists for various outdoor recreation activities in both the summer and winter. Survey information was obtained from visitors to the county on their trip to the area as well as how they would change their visitation with a decrease in the amount of ranchlands in the area. The potential change in consumer surplus was estimated for visitors to the county based on a decrease in the amount of ranchland in the county. Their results showed that 25 % of the sample would reduce visitation and 23 % would increase visitation with a decrease in the level of ranchland open space in the county. Overall, the results showed that there would be no net change in visitor consumer surplus in this area with a decrease in the amount of ranchlands in the county. An update to this study was completed by Ellingson et al. (2011) in the summer of 2005. The study compared the results of two different research methods: contingent valuation and contingent behavior. Using the CV methodology, the authors found a decrease in WTP of $129 per trip day if ranch lands were converted to urban uses. The percentage of respondents saying they would reduce their visitation to Routt County with a reduction in ranchland open space was 50 % in this sample, compared to 25 % in the earlier Routt County tourist study by Rosenberger and Loomis (1999). The study found the CV measure to be twice that of the CB estimate.

Other studies have been undertaken that estimate the value of ranchland open space in other mountain communities in Colorado. Orens and Seidl (2009) estimated the value of working landscapes to winter tourists in Gunnison County, Colorado. Gunnison County attracts many winter tourists is to ski areas in Crested Butte, and agricultural lands provide much of the scenic landscape on the drive to the ski areas. Information from tourist surveys was used in the estimation of a contingent behavior model to assess the value of working landscapes to winter tourists in the region. The survey results showed that a loss of all ranch open space would result in decreased visitation by winter tourists to Gunnison County. The median WTP per acre per visitor day to protect open space was estimated at $0.002, or $685 per acre for the entire winter tourist season (estimated for all winter tourists).

Cline and Seidl (2009) estimate the value summer tourists place on open space amenities of agricultural working landscapes in Chaffee County in the Rocky Mountains in central Colorado. Chaffee County draws many tourists during the summer months for activities such as whitewater rafting, hiking, camping, horseback riding, and off-road vehicle use, but is lacking many of the resort-type amenities available in the locations mentioned in the studies outlined above. This study used CB information from a survey of tourists to estimate the WTP of visitors to Chaffee County for changes in the level of open space, as well as potential related decreases in water quality. The estimation results showed a decrease in consumer surplus of $15 for a loss of 75 % of the agricultural open space in the county. The results show that an associated decrease in water quality would result in a substantially larger decrease in consumer surplus.

These studies show that both residents and tourists in the Rocky Mountains place considerable economic value on agricultural landscapes in the region. The results seem to indicate that these values may vary depending upon the location and type of site (resort area or more rural location). Additional research could be undertaken to provide more insight into how these values vary across different locations. Overall, the evidence indicates that there is some additional economic value placed on open space and other attributes of agricultural lands in the Rocky Mountains that are not currently being captured in the market.

5 Capturing Non-market Amenity Values

Several options exist that may allow farmers and ranchers to take advantage of the non-market values from agricultural working landscape amenities to help maintain the viability of their agricultural operation. These options can include taking advantage of agritourism or other niche businesses that may provide opportunities to capitalize on the non-market values tourists hold for agricultural working landscapes. Other policy options also exist such as conservation easements and other subsidy programs that may provide some financial assistance to farmers and help them to maintain working landscapes that provide amenity values to the community and to area visitors.

Agritourism, loosely defined as tourism related in some way to cultural, culinary or natural resource aspects of agriculture, has been considered more frequently in recent years as a way for farmers and ranchers to diversify their business (Wilson et al. 2006). Agritourism can include a wide range of different activities including outdoor recreation, educational activities, guest services, entertainment, and on-farm sales. Farm-based recreation activities have been developing throughout the United States, although a recent study by Brown and Reeder (2007) found both distance from a city and presence of natural amenities to be positively related to farmer involvement in a farm recreation business.

Farmers can benefit from agritourism not only through the additional income received but also as a way to diversify their business and help act as a cushion against uncertainty. Although other motivations may also drive farmers to consider expanding into agritourism ventures, economic motivations seem to dominate most decisions to expand. For example, Nickerson et al. (2001) found that Montana farmers most often noted economic motivations as their reason for entering into agritourism activities.

The remote nature and abundance of natural amenities in much of the Rocky Mountain region makes this area promising for the development of agritourism operations. In particular, wildlife-related recreation activities such as hunting, fishing, and wildlife viewing are often mentioned as having high potential in the region (Wilson et al. 2006; Coupal et al. 2004). Based on previous research that shows tourists to the region have a positive willingness to pay for agricultural amenities, it seems that significant potential exists for expansion of agritourism in the region.

Conservation easements are another option that has been exercised by many farmers and ranchers throughout the United States to maintain land in agricultural production while providing some financial incentives to the farmer that may assist in maintaining profitability. Conservation easements are voluntary, legally binding agreements between the land owner and another entity (often a non-profit land trust) that restrict future development and certain land uses in order to preserve societal benefits (natural and cultural amenities). Options exist for the landowner to sell the easement, to donate the easement and obtain tax benefits, or to engage in a hybrid of the two approaches (Keske et al. 2009).

Conservation easements and land trusts have been increasing in number in the Rocky Mountain States as the threat of development has grown in recent years. Albers et al. (2004) noted that Arizona, Colorado, New Mexico, and Utah had increased their protected area by 1,600 % in the previous 10 year period. Although land trusts have been popular throughout the region as a land conservation strategy, several studies have raised issues with the way markets are structured and noted specific areas of concern from landowners in the region.

A few recent studies in the Rocky Mountain region have assessed the attitudes of landowners with regard to their attitudes and experiences related to conservation easements. A recent study by Miller et al. (2010) found that most Wyoming and Colorado landowners surveyed were interested in conserving wildlife habitat, preserving open space, and maintaining land in agricultural production. Landowners believed that recreational opportunities were important to the general public, but felt that it would be important to be compensated for such use. Some of the main concerns of respondents included easements in perpetuity and requirements of public access (Miller et al. 2010). Marshall et al. (2002) found that Colorado landowners surveyed initially expected financial returns to be the main benefit from establishing conservation easements; however, benefits from land protection outcomes were more satisfying in the end.

Keske et al. (2009) also note three common market failures that affect markets for conservation easements: information failures, thin markets, and uncertainty. They suggest several reforms that may help to improve the efficiency of conservation easement markets including investments in government communications and research to reduce information failures, educating landowners and conservation organizations about each others' preferences to increase the depth of the market, making tax benefits progressive rather than regressive, and taking steps to reduce the uncertainty of future earning power. Providing better information and additional options (such as term easements) to landowners may help to make conservation easements a more feasible alternative for farmers in the region.

6 Conclusions and Policy Implications

Agriculture, particularly ranching and livestock production, has traditionally been important in the Rocky Mountain region, both economically and culturally. While the global outlook for livestock products is promising, certain other challenges exist for farmers in the region. In particular, the demand for non-agricultural land uses, and associated landscape change and land price effects are likely to continue to impact agricultural producers in the region. The abundant natural amenities in the Rocky Mountain region are likely to continue to draw amenity migrants and retirees to the area. This increased demand, along with the scarcity of privately-owned land in many counties in the Rocky Mountains, is likely to continue to put increasing pressure on farmers in the region to subdivide and sell their land for residential or commercial uses.

As shown in this chapter, a number of empirical studies conducted in the Rocky Mountain region have shown that there is considerable economic value embodied in these natural amenities, which continue to draw tourists and residents from other regions of the country. The non-market valuation estimates that have been calculated show that the "draw" of this region lies in the abundance of natural cover and the diversity of species that reside within that habitat. While this places pressure upon the land values as undeveloped areas are sought for conversion, it also provides an opportunity for local residents and farmers to capture some of this economic value.

A wide range of options exist that may allow local residents of this region to better capture the economic value that has been measured in the non-market valuation literature. Among these options are the development and promotion of agritourism operations that allow local producers to generate revenue from a wide range of visitors interested in the wildlife habitat and natural amenities supported by agriculture, on-farm educational and cultural opportunities, entertainment, and on-farm direct sales. Other options to help capture this value can include the sale or donation of conservation easements and associated tax benefits that come from those activities. Other state and federal assistance programs may also be combined with these activities to help enhance the profitability of agricultural operations. The specific set of options chosen by individual farmers will vary depending on many factors including the location of the farm, the preferences of the community and visitors to the area, and government assistance available in the state.

Federal, state, and local governments, as well as farmer outreach groups can provide farmers with information to determine the appropriate mix of available options that make the most sense for their farm. Additional training and support for agritourism ventures is an important area for expansion, since this is outside of the traditional area of expertise for both farmers and support groups and agencies. Furthermore, the markets for conservation easements should be reviewed to deal with inefficiencies. Addressing the reasons for non-participation will be important in moving these markets forward.

Farming and ranching will likely continue into the future in the Rocky Mountain region, however, specific issues stemming from land use and landscape change will need to be addressed. The natural amenities and public lands available throughout many areas of the Rocky Mountains make the area unique, but also create unique challenges. In order to maintain viable agricultural operations in the future, area farmers and ranchers will likely have to diversify and take advantage of new options such as agritourism, conservation easements, and government assistance alongside traditional farming and ranching activities.

References

Albers HJ, Ando AW, Kaffine D (2004) Land trusts in the United States: analyzing abundance. Resources 153:9–13

Bergstrom JC, Ready RC (2009) What have we learned from over 20 years of farmland amenity valuation research in North America? Rev Agric Econ 31(1):21–49

Bergstrom JC, Dillman BL, Stoll JR (1985) Public environmental amenity benefits of private land: the case of prime agricultural land. South J Agric Econ 17:139–149

Bowker JM, Didychuk DD (1994) Estimation of the nonmarket benefits of agricultural land retention in Eastern Canada. Agric Resour Econ Rev 23:218–225

Brown DM, Reeder RJ (2007) Farm-based recreation: a statistical profile. Economic research report No. 53. United States Department of Agriculture, Economic Research Service

Cline SA, Seidl A (2009) Surf and turf: tourists' values for multifunctional working landscapes and water quality in Colorado. Am J Agric Econ 91(5):1360–1367

Coupal R, Beauvais G, Feeney D, Lieske S (2004) The role and economic importance of private lands in providing habitat for Wyoming's big game. University of Wyoming Cooperative Extension Bulletin, No. B-1150. http://www.wyomingextension.org/agpubs/pubs/B1150.pdf

Diemer JA (1979) Agriculture, urban development, and the public domain in the Rocky Mountain West. Land Econ 55(4):532–536

Ellingson LJ, Seidl AF, Loomis JB (2011) Comparing tourists' behaviour and values of land use changes: a focus on ranch land open space in Colorado. J Environ Planning Manag 54(1):55–69

Geoghegan J (2002) The value of open spaces in residential land use. Land Use Policy 19:91–98

Gosnell H, Travis WR (2005) Ranchland ownership dynamics in the Rocky Mountain West. Rangel Ecol Manag 58(2):191–198

Gosnell H, Haggerty JH, Travis WR (2006) Ranchland ownership change in the Greater Yellowstone Ecosystem, 1990–2001: implications for conservation. Soc Nat Resour 19:743–758

Keske CM, Hoag DL, Bastian CT (2009) Can conservation easements market evolve from emerging to efficient? West Econ Forum 8(1):10–17

Le Goffe P (2000) Hedonic pricing of agriculture and forestry externalities. Environ Resour Econ 15:397–401

Magnan N, Seidl AF, Loomis JB (2012) Is resident valuation of ranch open space robust in a growing rural community? Evidence from the Rocky Mountains. Soc Nat Resour 25:852–867

Marshall A, Hoag D, Seidl A (2002) Colorado landowner conservation easement survey. Technical bulletin #02-3, Colorado Experiment Station, Colorado State University. http://www.colostate.edu/dept/aes/Pubs/pdf/tb02-3.pdf

Miller AD, Bastian CT, MacLeod DM, Keske CM, Hoag DL (2010) Factors impacting agricultural landowners' willingness to enter into conservation easements: a case study. Soc Nat Resour 24:65–74

Nickerson C, Hellerstein D (2007) Farmland preservation programs: another tool for managing urban growth? Amber Waves 5(2):6

Nickerson NP, Black RJ, McCool SF (2001) Agritourism: motivations behind farm/ranch business diversification. J Travel Res 40:19–26

Nickerson C, Ebel R, Borchers A, Carriazo F (2011) Major uses of land in the United States, 2007. EIB-89, U.S. Department of Agriculture, Economic Research Service. http://www.ers.usda.gov/media/188404/eib89_2_.pdf

OECD-FAO (2011) OECD-FAO Agricultural Outlook 2011. Chapter 7, Meat. http://www.oecd.org/dataoecd/2/36/48184304.pdf

Orens A, Seidl A (2009) Working lands and winter tourists in the Rocky Mountain West: a travel cost contingent behaviour and input-output analysis. Tour Econ 15(1):1–28

Ready RC, Berger MC, Blomquist GC (1997) Measuring amenity benefits from farmland: hedonic pricing vs contingent valuation. Growth Change 28:438–458

Riebsame WE, Gosnell H, Theobald DM (1996) Land use and landscape change in the Colorado Mountains I: theory, scale and pattern. Mt Res Dev 16(4):395–405

Rosenberger RS, Loomis JB (1999) The value of ranch open space to tourists: combining observed and contingent behavior data. Growth Change 30:366–383

Rosenberger RS, Walsh RG (1997) Nonmarket value of western valley ranchland using contingent valuation. J Agric Resour Econ 22(2):296–309

Seidl A (2006) Assessing the total economic value of ranching in mountain communities: an overview. Colorado State University Cooperative Extension, Economic Development Report, EDR 06-03

Vanslembrouck I, Van Huylenbroeck G, Van Meensel J (2005) Impact of agriculture on rural tourism: a hedonic pricing approach. J Agric Econ 56(1):17–30

Wilson J, Thilmany D, Sullins M (2006) Agritourism: a potential economic driver in the rural west. economic development report, EDR-06-01, Colorado State University Cooperative Extension

Conservation Agriculture in the Andean Highlands: Promise and Precautions

Jeffrey Alwang, George W. Norton, Victor Barrera and Ruben Botello

1 Introduction

In the Andean Region (AR) of South America, the challenge of increasing agricultural production while conserving or improving the natural environment has attained a sense of urgency. The AR is characterized by diverse natural, physical and social environments and this diversity contributes to the wide variety of farming systems observed across the region. While large areas of the higher-elevation areas have been cultivated under indigenous systems since pre-Columbian times, much of the region is farmed under smallholder systems which began to evolve following the major land reforms beginning in the Andean countries in the early 1950s. Prior to these reforms, the hacienda system predominated, extensive cultivation systems were wide-spread, and poor laborers were connected to the hacienda through institutional mechanisms such as indentured servitude. The reforms divided large areas into smallholder production systems, but most were not accompanied by provision of agricultural services such as extension or applied research. These post-reform areas comprise most of the Ecuador highlands and the Bolivian highlands outside the altiplano.

The post-reform AR is characterized by poor areas where productivity is low due to resource constraints such as small and shrinking farm sizes, poor soil quality, erratic rainfall, and exposure to natural risks (Alwang and Sowell 2010; Zimmerer 1993). Farmers are experiencing unprecedented stress due to natural

J. Alwang (✉) · G. W. Norton
Virginia Polytechnic Institute and State University (Virginia Tech), Blacksburg, USA
e-mail: alwangj@vt.edu

V. Barrera
Instituto Nacional Autónomo de Investigaciones Agropecuarias INIAP, Quito, Ecuador

R. Botello
Fundacion PROINPA, Cochabamba, Bolivia

S. Mann (ed.), *The Future of Mountain Agriculture*, Springer Geography,
DOI: 10.1007/978-3-642-33584-6_3, © Springer-Verlag Berlin Heidelberg 2013

factors such as climate change which reduces water availability, increases variability of rainfall and temperatures, and contracts the spread of rainfall events and, hence, the planting season. Despite the undeniable importance of climate change, an equal or bigger stress is coming through economic integration of these areas with the broader national and global economies. Market forces are rapidly shaping these areas, reducing labor availability, increasing access to alternative inputs, making information more widespread, and raising the economic values of agricultural products (Amaya and Alwang 2012). In many AR communities, outmigration of workers in search of improved incomes is driving up the price of labor. As labor becomes scarce, farmers seek alternatives such as new, less labor-intensive crops, or new management practices such as increased mechanization. These factors are leading to changes in farm structure (Amaya and Alwang 2012). The economic transformation affecting agriculture elsewhere in the developing world is beginning to impact mountain farming systems in the AR.

Settled populations in the AR have adapted to environmental challenges by adopting low-input production practices, expanding the agricultural frontier into fragile areas, and diversifying their livelihoods (Larochelle and Alwang 2012). While AR livelihoods have always been diverse, growing opportunities in cities and even other countries are increasing this diversity (Andrade 2008; Lanjouw 1999). Penetration of public services such as water, electricity, schooling and health care creates demand for money incomes to pay for these services; this demand further stimulates market integration. Integration has also raised the demand for new agricultural technologies; outmigration of labor increases demands for labor saving technologies, while market prices and their fluctuations are creating incentives for new production practices and new crop alternatives (Carrion 2013). In Bolivia and Ecuador, farmers in areas that traditionally specialized in potato production are increasingly diversifying into high-valued products such as quinoa, or production systems that provide more stability with less labor such as dairy.

These stresses require new and rapid adaptations on the part of farm households in the AR; system resilience to increased vulnerability is of critical concern. Highland yields are already low, crops are nutrient deficient, soil structure is poor, and erosion is widespread and severe. Inadequate and unpredictable rainfall means crop production is threatened by chronic soil moisture stress, and intense, short-duration precipitation contributes to runoff. AR farms are often dependent on a single crop—the potato—and regional food insecurity can be reduced through improvements in potato yields, reduced yield variability, and introduction of companion crops to raise incomes. While new technologies may be necessary, they are rarely a sufficient means of improving long-term well-being. Investments in household assets can complement such technologies and assist households with their integration into rapidly changing markets. These investments can help provide a means of escape from long-term poverty traps.

This chapter examines the potential of conservation agriculture (CA) to increase agricultural productivity and sustainability in highly sloped areas and inter-Andean valley in post-land reform areas of the AR. Conservation agriculture (CA) is as a set of practices intended to conserve, improve and make more efficient use of resources through integrated management of soil, water and biological resources combined with external inputs. In its purest form, CA is based on three principles: (i) minimum or no mechanical soil disturbance; (ii) permanent organic soil cover; and (iii) diversified crop rotations. The chapter summarizes existing literature about the potential for CA in the AR highlands, with a close focus on projects in Ecuador and Bolivia where the smallholder model of CA is being tested and research results are available.

The chapter critically evaluates the potential of CA and treats it as a means, not an end. The end concerns of efforts to promote CA are improved household well-being, enhanced food security, higher agricultural productivity, and environmental sustainability. It is imperative to address the many constraints to addressing these concerns, and these include pathogens and diseases in the agricultural system, off-farm barriers to higher income activities, and institutional constraints that impede the ability of farmers to capture values from off-farm environmental improvements such as the values of sequestered carbon and reduced siltation in rivers.

2 Conservation Agriculture: Potential and Promise

Conservation agriculture may present a potential solution to lagging productivity and environmental degradation in the AR. It also offers solutions to increasing labor costs and can help farmers manage declining availability and increased irregularity of soil moisture. Conservation agriculture is an environment-focused approach involving farming practices aimed at conserving soil and improving soil health and productivity. The three main principles of CA are use of reduced or minimum tillage, maintenance of soil cover, and adoption of improved crop rotations (FAO 2008). These practices build soil health over time by improving organic content, enhancing moisture retention and water holding capacity, fostering growth and preservation of soil micro-organisms and efficiently using available nutrients. Income growth in CA can result from higher productivity, reduced use of purchased inputs such as herbicides and labor to manage weeds, and gradual declines in production costs as labor is replaced with mechanized processes.

A primary goal of CA is to increase soil health as measured by organic content, especially carbon, in agricultural soils. Higher organic content is linked to greater soil productivity, production of higher-nutrient food crops, and increased carbon sequestration (Shaxson et al. 2008; West and Post 2002; Lal 1997). While the bulk of the CA literature focuses on on-site benefits, CA can reduce off-farm damages from erosion (Kassam and Friedich 2011).

Literature on benefits of CA has examined changes in yields and farm income, the impacts on the environment and the natural resource base, the profitability and success of individual CA components, and factors affecting adoption (for an overview of the spread and implications of conservation agriculture in the Americas, see Derpsch 2005). Much of the documented successes in CA have come from extensive systems (e.g. maize and wheat in the developed world, soybeans in South America) or relatively small-scale systems under intensive farming in flat areas (e.g. the rice–wheat system in the Indo-Gangenic plains; see Hobbs et al. 2006). Research on CA systems for smallholders has focused on Sub-Saharan Africa, and generally indicates positive environmental and economic impacts (Jenrich 2011; Marongwe et al. 2011).

Despite evidence that the Mayans used CA techniques in the Central American highlands by using sticks to insert maize seeds in unprepared soil, CA has not gained substantial penetration into highland production systems in Latin America. Several studies have demonstrated that individual conservation practices can be profitable in highland AR systems (Swinton 2000; Swinton and Quiroz 2002; Knowler and Bradshaw 2007), few studies have looked specifically at highland AR smallholder systems. While AR cropping systems are remarkably diverse, a broad division can be made between traditional systems (mainly found in indigenous areas of the altiplano) and post land reform systems. The latter represents farming system on lands that were originally held in extensive production systems but have evolved following the break-up of the large haciendas in the mid-20th century. We focus on examples from the post land reform areas.

3 Post Land Reform Areas of the Andean Region

Much of the Ecuador and Bolivian Highlands were converted into large-scale holdings following the European invasion. The Chimbo River sub-watershed in Ecuador and the Tiraque area of Bolivia are broadly representative of post-reform conditions in the AR. They vary in that soil conditions are better and rainfall is more abundant in Chimbo compared to Bolivia, but these differences help illustrate how the general principles of CA can be useful under a wide variety of environments. The areas also have hosted ongoing conservation agriculture experiments as a part of SANREM CRSP long-term research.[1]

3.1 Chimbo River Watershed in Ecuador

The Chimbo River sub-watershed in south-central Ecuador covers approximately 3,635 km^2 (Fig. 1). A program for research on CA in the area was established in 2008 in two micro-watersheds of the Chimbo: the Illangama and Alumbre.

[1] The research was funded by USAID through the Sustainable Agriculture and Natural Resource Management Collaborative Research Support Project (SANREM-CRSP). This research was made possible by the United States Agency for International Development and the generous support of the American people through USAID Cooperative Agreement No. EPP-A-00-04-00013-00.

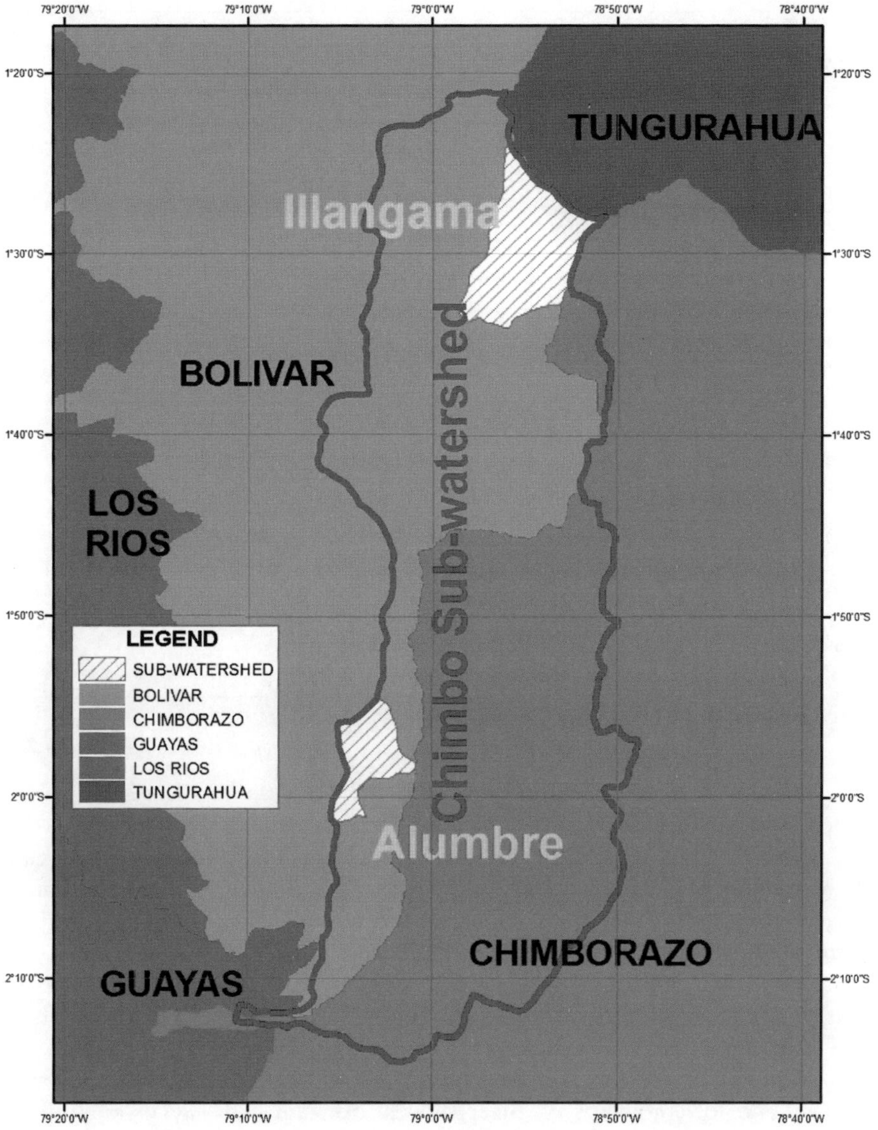

Fig. 1 The Chimbo River area of Ecuador

The Illangama micro-watershed covers 131 km^2 and extends from a latitude of 1°23′55.30″S through 1°34′4.80″S and from 78°50′39.38″W to 78°58′29.52″W. The Alumbre covers 65 km^2 and extends from 1°54′29.14″S to 2°1′36.90″S and from 79°0′22.20″W to 79°6′4.41″W. The Illangama is between 2,800 and 4,500 masl,

Table 1 Conditions in Chimbo sub-watershed

Agro-ecological conditions	Productive activities
Illangama	
Region: Páramo and Andean mesa	Agriculture—potato (*Solanum tuberosum*), pasture, quinoa (*Chenopodium quinoa*), faba (*Vicia faba*), chocho (*Lupinus mutabilis*) and barley (*Hordeum vulgare*)
Life zones: Subalpine or Boreal, Montain, Low Mountain and Cool Temperate	Animal production—cattle and others
Temperature: 7–13 °C	Tourism, artisan production, dairy and off-farm labor supply
Altitude: 2,800–5,000 m	
Annual rainfall: 500–1,300 mm	
Alumbre	
Region: Andean mesa and subtropical	Agriculture—maize (*Zea mays*), beans (*Phaseolus vulgaris*), peas (*Pisum sativum*), blackberry (*Rubus glaucus*), tree tomato (*Ciphomandrea betacea*), vine tomatoes (*Lycopersicum esculentum*)
Life zone: Low Mountain and Pre Mountain	Animal production—poultry, swine
Temperature: 15–19 °C	Agro-industry-medicinal plants, cacao (*Theobroma cacao*), organic coffee (*Coffea arabica*)
Altitude: 2,000–2,800 m	Off-farm labor supply
Annual rainfall: 750–1,400 mm	

Source INIAP-SANREM CRSP-SENACYT, 2006

with agricultural activity found between 2,800 and 3,600 masl. The Alumbre area ranges from 2,000 to 2,800 masl with agriculture throughout (INIAP 2008).

The watersheds are characterized by social and economic conditions that threaten environmental sustainability and create long-term risks to human populations (Barrera et al. 2008). The area is among the poorest in Ecuador (INEC-MAG 2002). Water quantity and quality have declined in recent years, partly due to soil erosion, deforestation and expansion of the agricultural frontier into fragile highland páramo areas, which are reservoirs of clean water (GPB 2004; Gallardo 2000). Households depend on agriculture; more than 60 % of the economically active population in Bolivar Province is dedicated to agriculture. Agriculture is characterized by small holdings and low productivity. Steep slopes, irregular and sudden rainfall, and infrequent use of cover crops and other means of conserving soils are associated with severe soil erosion; particularly in Alumbre, clear signs of extreme loss of soil are visible. The Illangama is dominated by a potato-pasture production system, while the Alumbre is characterized by a mixed maize-beans system with minor grains and other crops (Table 1). The Instituto Nacional Autonomo de Investigaciones Agropecuarias (INIAP—the national autonomous agricultural research institute) has worked for many years in the Chimbo area and has built up an impressive network of collaborators in the area.

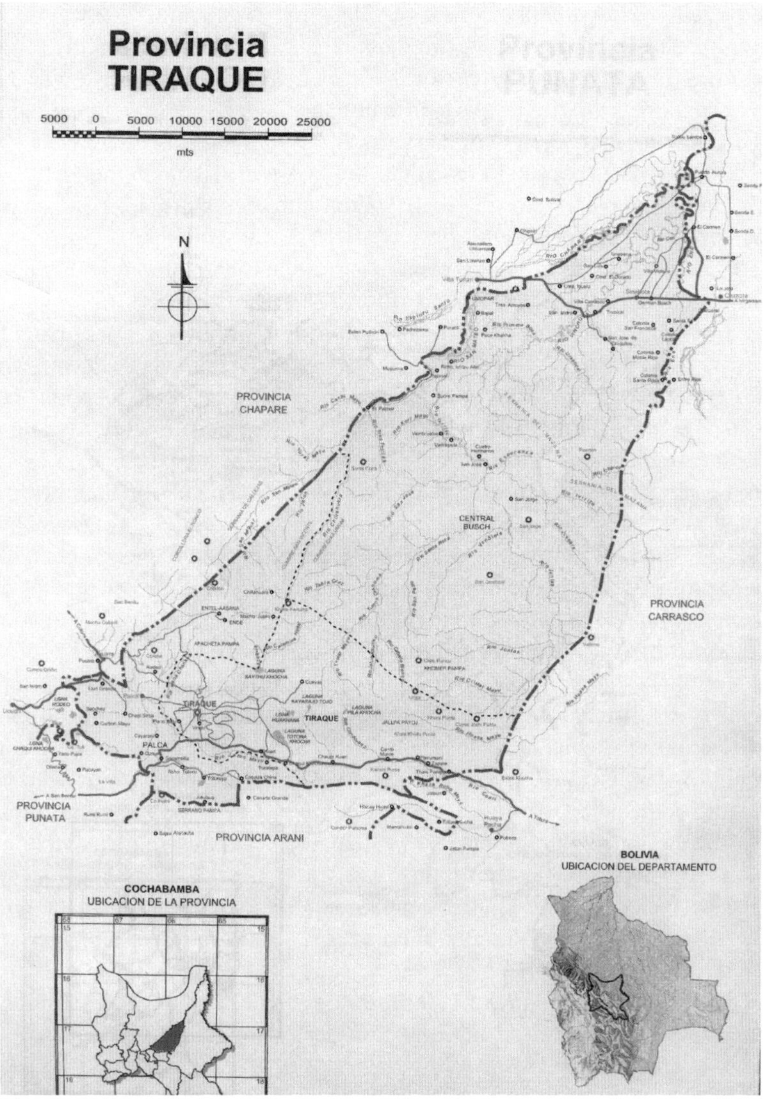

Fig. 2 The Tiraque area of Bolivia

3.2 Tiraque, Bolivia

The study site in Bolivia is located in the Jatun Mayu watershed, southeast of Tiraque municipality, approximately 75 km. from Cochabamba (Fig. 2). The watershed covers approximately 117 Km^2, and is between 3180 and 4300 masl. While the slopes in the area are not as steep as in the Chimbo area, they are between 10 and 25 % on cultivated fields and between 20 and 40 % in pastured areas. The area is characterized by low-quality soils and relatively low rainfall (between 650 and 850 mm annually). Farmers in the area note that rainfall is increasingly erratic and concentrated into ever shorter periods. Canal irrigation systems, however, allow farmers with access to water rights to produce small quantities of potatoes in the drier periods following the main potato harvest.

The roughly 4000 people in the Jatun Mayu are distributed among 14 communities. Landholding sizes are relatively small, with the vast majority of households owning between 1 and 5 hectares (Amaya 2009). Many of these families are direct descendants of *colonos* (workers) from the large-scale Torolapa and other haciendas predominating before the 1953 land reform. Prior to the land reform, hacienda owners allocated *colonos* small plots of land near their houses. These parcels were farmed as sharecroppers with the hacienda owners; following the reform these families gradually gained title to fragmented parcels in scattered locations near and fairly distant from the homestead (Dorsey 1975; Larochelle and Alwang 2012). Despite the small holding sizes, household incomes are dependent on agriculture although sources of income have diversified substantially in recent years (Amaya 2009). The most important crop in the area is potatoes, followed by small cereals, beans and, recently quinoa. Agricultural production is concentrated between October and April during the rainy season, with secondary (mostly potato) production occurring under irrigation during the drier months. Production is labor-intensive, with little mechanization and purchased agro-chemicals. Most output is consumed by the household, but between 20 and 50 % is sold (Amaya 2009).

Social problems are related to low levels of income and lack of income-earning alternatives. Recent evidence shows out-migration to cities and, more frequently, the humid lowlands is a common strategy. Some migration is seasonal, but much is permanent, contributing to labor shortages and increasingly diversified income sources. Environmental problems are related to farming: soil erosion, low and declining soil fertility, and changing rainfall patterns are frequently cited as concerns by residents. The Fundacion para la Promocion e Investigacion de Productos Andinos (PROINPA—the Foundation for Promotion and Research of Andean Products) has worked in Tiraque since its establishment in 1989. PROINPA researchers have engaged local farming sindicatos[2] in establishing farmer-led research projects to address production constraints.

3.3 Common Challenges

Farmers in these and other post land-reform areas face environmental and economic stress. Weather and climate factors induce high levels of risk, while economic forces such as market integration and outmigration of labor exacerbate these risks.

---

[2] The sindicato is the basic unit of organization for agricultural producers in the inter-Andean valleys of Bolivia.

Risk management has been a central component of farming operations. Decision makers diversify their farming operations and spread activities over space to self-insure against idiosyncratic risks (Dorsey 1975; Larochelle and Alwang 2012; Castelhano et al. 2012). They diversify income sources by sending migrants and engaging in a multiple income-earning activities (Andrade 2008; Lanjouw 1999; Reardon et al. 2001). While diversification has helped them manage many risks, farmers now demand new cropping alternatives and technologies to increase yields and lower costs of production. Thus, while risk management is still (and perhaps more) important, secular trends in soil productivity (downward), crop prices (generally upward), and input costs (particularly labor, which is becoming increasingly costly) are stimulating demands for new farming practices (Amaya and Alwang 2012; Barrera et al. 2008; SANREM 2012). Farming practices which evolved slowly over time are now progressing at a rapid rate.

3.4 Challenges to Conservation Agriculture in the Andean Region

Conservation agriculture has potential to assist AR farmers in adapting to the stress they face. Conservation agriculture is found throughout the world; estimates are that more than 80 million hectares are farmed using CA techniques (Garcia-Torres et al. 2003).[3] Developing-country CA gained momentum in South America during the 1980s with wide adoption of reduced and no-tillage soybeans in Paraguay and Brazil. Most CA adoption in South America has, however, occurred in low-slope fields and under mechanized conditions. CA has not been widely adopted in the AR, despite the fact that lands in the AR are among the most fragile and productivity-challenged in the world (Winters et al. 2004). Part of the reason for lack of widespread use of CA is related to AR farming which has characteristics creating special challenges to CA. Additionally, institutional factors, particularly the absence of publicly funded agricultural extension, have limited the spread of CA in the AR.

Important factors limiting adoption of CA in developing country agriculture include potential increases in input costs in the short run, competing uses for ground cover crops, and limited access to the necessary inputs. In the short run, as CA practices are being learned by the farmer, costs can be higher and yields lower, and necessary investments in farming implements may increase the short-run burden of CA (Govaerts et al. 2009; Giller et al. 2009). Organic cover crops, which form an important part of CA farming, are often highly valued by smallholder farmers as feed for ruminants; convincing them that the crop provide more value as a cover can be a challenge. No-till farming implements such as direct seeders

[3] Exact estimates are difficult to come by because the criteria used to identify conservation agriculture vary from study to study. It is estimated that more than 100 million hectares worldwide are currently planted under no-till.

often form a critical part of a CA system; often these implements are not available at a scale suitable for small holding sizes, and many small-scale farmers lack resources to pay for them even when they are available. These challenges are common to many regions of the developing world where CA adoption has lagged behind its potential (Knowler and Bradshaw 2007).

In addition to these factors, conditions in the AR present more specific obstacles.

Small holding sizes and diverse conditions in densely populated areas. Densely-populated, intensively-cropped AR areas are characterized by small holding sizes and production systems with remarkable diversity (Barrera et al. 2008). Worldwide, CA is predominantly found in relatively flat areas where topography is homogeneous (FAO 2006). The mountainous topography, characterized by small and fragmented plots makes CA more of a challenge in the AR. Technologies need to be developed and tested to ensure that they are suitable for these conditions. For example, direct seeding implements for quinoa are widely used in the altiplano where extensive planting is common, but these same implements are inappropriate on undulating and relatively small fields. Similar technologies need to be developed for these specific conditions.

Multiple cropping. In most of the AR, smallholders practice multiple cropping, either by growing two or more crops in sequence on the same field each year, or by intercropping two or more crops on the same field. Sequential cropping systems, especially on steep slopes, can degrade soils because little or no time is devoted to restoring soil health between periods of crop production. Intercropping can be compatible with maintaining ground cover; for example, in Ecuador bush beans are planted in established pastures and cover created by the pasture reduces soil and nutrient losses. However, intercropping can create special challenges to CA as all intercropped varieties should be planted to reduced tillage. Seed depths are likely to vary and few direct seeders are compatible with intercropping.

Crop mix. The potato is the centerpiece of the Andean diet; as much as 50 % of dietary calories come from it. Andean potato cultivation usually involves intensive tillage. During soil preparation, the entire topsoil is loosened and pulverized into small aggregates. Some CA techniques exist for potatoes, including mulch planting in combination with green manure and even no-till on smallholder farms. In Peru, farmers press seed potatoes into the ground and groundcover is maintained by mulch or plastic (Bunch 1999). In Colombia, farmers have adopted a mulch-based potato conservation agriculture production system (see Fig. 3 for examples from Bolivia). Thus, there is hope for CA even in potato-based agriculture. There is evidence that deep-rooted cover crops can play a role in a potato CA program, helping suppress soil pathogens and weed populations (McGinnis 2006). Throughout the AR, however, variable potato prices, high dependence on hired labor and increasing pest pressures are leading farmers to diversify into alternatives (Carrion 2013).

In Honduras, research has demonstrated that corn can be hand-sown into the residue of a preceding legume cover crop such as *Mucuna* or *Lablab* spp., with legume residue providing both crop fertility as well as a high degree of weed

Fig. 3 Direct seeded potatoes in Bolivia

suppression (Melara and del Rio 1994). Similar 'slash-mulch' systems have been developed with some success in Peru (Iberico 1994).

Quinoa is a protein rich crop that is increasingly important for the Andean diet. Increasing world demand has led to increased production for export in recent years and spread of production out of the altiplano into sloped areas and inter-Andean valleys where post land-reform agrarian structures predominate. As a small grain, quinoa is amenable to direct seeding (in much of the altiplano, quinoa is directly seeded), but small holding sizes outside the altiplano make altiplano-suited direct seeders inappropriate. Quinoa production has become common in our study areas and research on CA rotations including quinoa is advanced.

Livelihoods and poverty. Andean smallholders have diversified livelihood strategies, and these strategies could interfere with adoption of CA (Giller et al. 2009). Households engage in many income-generating activities on and off the farm, these strategies are often time intensive and CA practices need to fit into existing livelihood systems (Shiferaw et al. 2009). Efforts to introduce new production practices must consider multiple factors in household decision making. High poverty and few assets among AR smallholders mean that challenges such as access to finance to fund new equipment and input purchases must be overcome (Andrade 2008; Amaya 2009).

Establishment of ground cover and competing uses. Andean agriculture is predominantly rain-fed with rainfall concentrated in a short period once a year.

Some AR highland producers have access to supplementary irrigation through canal-based irrigation systems, and those that do often avail of a second cash crop (usually potatoes). Following harvest of the primary crop, rainfall is minimum, often insufficient to establish a cover crop. Obviously, establishment of such a crop implies competition with cereal or potato crops.

In temperate corn-bean-wheat systems, for example, vetch or clover spp. are planted after wheat harvest, and achieve most growth during the subsequent fall and spring prior to the corn crop. In this case, the cover crop does not conflict with the growing season for the primary grain crops. In much of the AR, however, the optimum establishment and growth periods for the cover crop is similar to that of the target food or cash crops. One alternative is to employ an intensive fallow system where a growing season is dedicated to production of the cover crop or where a cover crop is intercropped with the food or cash crop. An alternative, which we have employed, is to avail of existing soil moisture during the post-harvest period and supplement, where possible, with irrigation.

Experimentation with such techniques has demonstrated the potential and limitations of cover crop establishment in CA systems. In our study areas, cover crops are never accepted by growers as an alternative to their cash crop (whether potatoes, maize and beans or quinoa); they need to be grown off season. In areas with enough residual moisture or intermittent off-season rainfall, an oats-vetch mix can be relatively rapidly established without supplemental irrigation. But in approximately 30 % of the trials we have run, cover crops cannot be established without additional moisture. The issue then becomes the value of competing uses of the water. In our context, which is typical of many AR environments, smallholders own shares (*acciones*) in the canal system. Each share allows them a certain flow of water at designated times. The value of these shares was included in our evaluations of the CA system, which thus included the full cost of establishing cover.

Institutional challenges. Widespread adoption of CA in the Andean Region is constrained by institutional challenges some of which are widespread in the developing world. Smallholders everywhere would receive additional benefits from CA if they could capture external benefits from their on-farm actions. Benefits from CA-related carbon sequestration accrue to everyone and markets currently exist for sequestered carbon. Unfortunately the Clean Development Mechanism (CDM) of the Kyoto protocol excludes soil carbon sequestration. In addition, monitoring and administration costs for certifying soil carbon sequestration benefits for regulatory markets (such as the European Union Emissions Trading Scheme) present an important obstacle for entry by smallholders into these markets. In sloped areas, an additional external benefit is felt through reduced off-farm damages from erosion. In Ecuador, for instance, downstream flooding from siltation in the coastal plains is associated with millions of dollars in damages every year (Barrera et al. 2008). Programs for payments for environmental services to smallholders are constrained by high monitoring and administration costs.

Perhaps the most important institutional obstacle to widespread adoption of CA in the AR is the absence of publicly funded agricultural extension in the region. Ecuador and Bolivia's extension services were effectively eliminated during the

1990s and the limited outreach that occurs comes through research and other projects, NGOs, and other targeted, limited-duration efforts. Because there is little potential for private-sector promotion of CA among smallholders, diffusion of successful technologies will be slow and painstaking.

4 Evidence from Ecuador and Bolivia

Beginning in 2007 in Ecuador and 2009 in Bolivia, CA system trials were established in the research sites.[4] Field studies consisted of a randomized factorial design of: (1) cover crop or no cover and (2) conventional tillage or no/reduced-tillage. Rotations suitable to farmers were incorporated into this design (Table 2). Evaluation of the treatments included parameters measuring soil quality, including organic matter and water-holding capacity, soil erosion, productivity and costs of production of the different practices. Complementary satellite experiments dealt with growth constraints in a phosphorus-poor soil, the effects of access to irrigation during droughts, use of bacillus spp. and other biological inputs to stimulate growth and combat pests and diseases. Research on direct seeding and reduced tillage planters for quinoa and maize systems was also undertaken in participation with agricultural engineering departments in partner universities. Participatory socioeconomic research was used to gauge the acceptability of cover crops, new potato and other varieties, and farmer impressions of alternative CA practices.

The research on CA in Ecuador and Bolivia have been underway long enough to make general observations. First, over the relatively short periods under evaluation (4 and 3 years) the CA practices have yielded the same or slightly more as the farmer practices, while costs of production in the CA system are the same or lower than conventional practices. The differences are small, but in the near term, the CA systems are at least as good as conventional practices. Second, again in the short run, indicators of soil health are improving under the CA system. Soil chemical analysis of N, P and K, pH, and percent of soil organic material were conducted and, particularly in the case of organic matter, improvements are being observed over time. Physical measures of compaction, bulk density and soil moisture content are also being taken and the CA trials show slight improvements (statistically significant in the context of moisture content). Third, the results from the two prior observations indicate that, over time, the CA trials will perform at least as well and likely much better than the conventional practices. In fact, the last year of data showed rates of return to CA exceeding 10 %.

Third, farmers in the area are reluctant to adopt the practices. The project has conducted a number of farmer field days and other outreach efforts to engage farmers. The experiments are being run by organized farmer groups. Despite these

---

[4] Research partners were INIAP in Ecuador and PROINPA in Bolivia. Local governmental and non-governmental organizations were also included in the research program.

Table 2 Description of conservation agriculture experiments, Ecuador and Bolivia

Ecuador Illangama sub-watershed: potato–pasture system	Alumbre sub-watershed: maize–beans system	Bolivia Tiraque trials
Tillage: conventional, reduced and direct seeding	Tillage: conventional, reduced and direct seeding	Tillage: conventional, reduced and direct seeding (when posible)
Rotations: (i) Ground cover (dry season), potato, barley, faba, and forage mix; (ii) Potato, barley, oats-vetch and forage mix	Rotations: (i) Ground cover(dry season), hard maize, bush beans, hard maize, peas and hard maize; (ii) Hard maize, bush beans, hard maize, oats-vetch and hard maize	Rotations: (i) Ground cover (dry season), potato, faba; (ii)) Ground cover (dry season), potato, quinoa
Management: Ground cover use (fallow, grass with residuals removed, grass with residuals retained), fertilization with N and cover crops (faba and quinoa)	Management: Cover crops (peas, oats-vetch, maralfalfa–*Pennisetum sp.* and native trees)—incorporated/not into soil; fertilization	Management: Ground cover use (fallow, grass with residuals removed, grass with residuals retained), barley as a mulch bed for potato, fertilization with poultry litter, use of bacillus
Intensive pasture management (improved forages) with overseeding of clover	Use of maralfalfa and fruit trees in contours to form live barriers	
Erosion-reduction measures: deviation ditches with milín grass (*Phalaris tuberosa*) and native species; live barriers with native species	Erosion-reduction measures: live barriers with native species; fruit trees on contours to form live barriers (blackberry and narajilla); contour planting with alfalfa (*Pennisetum sp.*) in strips	Erosion-reduction measures: live barriers with native species; contour planting in strips

Source INIAP-SANREM CRSP-SENACYT, 2009

efforts to obtain farmer acceptance, suspicion lingers about the ultimate profitability and risks associated with CA. The concept of leaving crop residues on the ground is exotic, and the idea of not cultivating the soil to plant potato seeds is even more so. Farmers say they need more evidence of profitability of CA before they adopt. Direct seeded grains such as quinoa and maize have gained more acceptance. Sales of direct seeders for quinoa are robust in our Bolivia study area and, as a result, costs have begun to decline.

Fourth, the biggest unknown to date in the CA management system is treatment of the cover crop. Low costs of canal irrigation water in the areas have led to fairly wide planting of off-season covers. In areas without irrigation, the risk of cover crop failure in the absence of some off-season rains is high. In all areas, use of the cover crop is contentious. We have determined that herbicides (as opposed to hand cutting) is the preferred means of killing the crop, but farmers are conflicted about its use as feed or as ground cover. Over time, soil health will grow with continuous ground cover, but the productivity gains need to be substantial before farmers are willing to sacrifice the feed value.

Fifth, farmers are convinced that pressures for moisture- and labor-saving technologies will only grow over time. Labor costs have grown by more than 50 % since project inception and these increases are not slowing. These pressures (together with the increasing prices of quinoa and dairy products) will continue to push AR farmers toward CA-type technologies.

Sixth, while farmers recognize that their on-farm practices have off-farm consequences, they believe that the ability to capture benefits from CA practices is too far into the future to be a factor in current decisions. Our CA trial show substantially less soil erosion and soil organic content is clearly greater, but farmers place less weight on these factors than on additional profits and household food security. This is a clear disadvantage to CA-short-term concerns outweigh long-term benefits, especially for poor farmers in credit-constrained environments.

5 Conclusions

Stressors affecting farmers in the Andean highlands of South America require continuous adaptation to rapidly changing conditions. Economic stressors include integration into a growing regional and global economy and consequent changes in factor and product prices. Environmental stressors include shorter and more violent periods of rainfall, declining soil fertility, and reduced access to water for farming. Changes resulting from adaptation to these stressors are already visible in the region as farmers have diversified income sources, are changing crop and management mixes in response to market signals, and experimenting with new technologies to deal with environmental stress. Demand is high for options to facilitate this adaptation.

While conservation agriculture is not the answer to all farmer needs in the AR, it does represent a promising alternative for many, particularly in areas where post land-reform agrarian structures predominate. While CA has been widely adopted throughout the globe in extensive systems, adoption has lagged among smallholders, but its promise is slowly emerging. In the AR, important obstacles to CA exist, but evidence suggests that despite these obstacles, certain systems can work. Under some conditions CA implies no higher costs and, in the short term, produces equivalent or greater yields than conventional alternatives. Over time, its advantage will grow as economic forces make CA inputs more readily available and conventional inputs, particularly labor, more costly. Over time as well, soil health improvements under CA will raise productivity, making CA a long-run improvement for many farmers.

Institutional changes will help; these include enhanced outreach for smallholders, and better access to credit. Access to credit will facilitate equipment purchases, assist in risk management, and to facilitate longer-term outlooks that are more compatible with conservation agricultural approaches (e.g. Alwang et al. 2001). Larger institutional changes are needed to allow smallholders to capture the external benefits from CA adoption, and it is not likely that these institutional changes are imminent in the AR.

The writing is on the wall for smallholders in the mountainous regions of South America: continued change is likely to be the only constant over the next decade. Farmers will be forced to adapt to external stressors and need tools to assist in this adaptation. Conservation agriculture is one tool, but many kinks still need to be worked out, particularly given the combination of conditions facing AR farmers. The evidence presented in this chapter suggests a wider role for CA, but the concept needs to be tested area by area. Heterogeneous conditions preclude identification of a single solution, but CA principles hold promise across many diverse areas. Andean region farmers will adapt—and it is likely that 10 years hence farm structures and modes of operation will be markedly different than they are now.

References

Alwang J, Sowell A (2010) Socioeconomic factors affecting soil and water conservation in South America. In: Napier T (ed) Human dimensions of soil and water conservation. Nova Science Publishers, Hauppauge

Alwang J, Siegel PB, Jorgensen S (2001) Vulnerability: a view from different disciplines. SP Discussion Series 0115, Human Development Network, Social Protection Unit. The World Bank, Washington

Amaya N (2009) Effects of access to information on farmer's market channel choice: the case of potato in Tiraque Sub-watershed (Cochabamba, Bolivia). MS Thesis, Virginia Polytechnic Institute and State University, Blacksburg, VA (Unpublished)

Amaya N, Alwang J (2012) Women rule: potato markets and access to information in the Bolivian highlands. Agric Econ 43(4):403–413

Andrade R (2008) Livelihood strategies of farmers in Bolivar, Ecuador: asset distribution, activity selection and income generation decisions in rural households. MS Thesis, Virginia Polytechnic Institute and State University, Blacksburg (Unpublished)

Barrera V, Alwang J, Cruz E (2008) Manejo integrado de los recursos naturales para agricultura de pequeña escala en la subcuenca del río Chimbo—Ecuador: aprendizajes y enseñanzas, INIAP–SANREMCRSP–SENACYT, Boletín Divulgativo No. 339. Quito, Ecuador

Bunch R (1999) Reasons for the non-adoption of soil conservation technologies and how to overcome them. Mt Res Dev 19(3):213–220

Carrion V (2013) Impacts of IPM CRSP research on potatoes in Carchi Ecuador. MS Thesis, Virginia Polytechnic Institute and State University, Blacksburg (Unpublished)

Castelhano MJ, Alwang N, Kuminoff, Botello R (2012) Do farmers choose to produce potato varieties as a means to better manage risk? Revista de Agricultura (Bolivia) 51:39–46

Derpsch R (2005) The extent of conservation agriculture adoption worldwide: implications and impact. Proceedings of the third international congress of conservation agriculture, Nairobi, Kenya

Dorsey JF (1975) A case study of ex-hacienda Toralapa in the Tiraque region of the upper Cochabamba Valley. Land Tenure Center, University of Wisconsin, Madison

FAO (2006) Conservation agriculture web-site. www.fao.org/ag/ca

FAO (2008) International year of the potato: potato and soil conservation. Mimeo http://www.potato2008.org/pdf/IYP-8en.pdf

Gallardo G (2000) Informe Final Memoria Técnica Programa de Manejo Integrado de Recursos Naturales en Cuencas Hidrográficas y un Plan de Inversiones, en el Sector Agropecuario, MAG-BID-IICA. Quito, Ecuador

Garcia-Torres L, Benites J, Martinez-Vilela A, Holgado-Cabrera A (2003) Conservation agriculture: environment, farmers experience, innovations, socio-economy, policy. Kluwer Academic Press, Boston

Giller K, Witter E, Corbeels M, Tittonell P (2009) Conservation agriculture and smallholder farming in Africa: the heretics' view. Field Crops Res 114:23–34

Govaerts B, Verhulst N, Castellanos-Navarrete A, Sayre K, Dixon J, Dendooven L (2009) Conservation agriculture and soil carbon sequestration: between myth and farmer reality. Crit Rev Plant Sci 28(3):97–122

Hobbs PR, Gupta R, Meisner C (2006) Conservation agriculture and its applications in South Asia. In: Uphoff N, Ball AS, Fernandes E, Herren H, Husson O, Laing M, Palm C, Pretty J, Sanchez P, Sanginga N, Thies J (eds) Biological approaches to sustainable soil systems. Taylor and Francis, Boca Raton

Iberico (1994) A traditional slash/mulch system for beans in Peru. In: Thurston D, Smith M, Abawi G, Kearl S (eds) Slash/Mulch: how farmers use it and what researchers know about it. CATIE-CIIFAD, Ithaca

INEC-MAG, III Censo Nacional Agropecuaria: Resultados Nacionales, Provinciales y Cantonales. Instituto Nacional de Estadísticas y Censos y Ministerio de Agricultura y Ganadería, volumen 1. Quito, Ecuador, 2002

INIAP, Sistema de Información Geográfica de la Subcuenca del río Chimbo, Bolívar-Ecuador. Red de Monitoreo Climático. Instituto Nacional Autónomo de Investigaciones Agropecuarias, 2008

Jenrich M (2011) Potential of precision conservation agriculture as a means of increasing productivity and incomes for smallholder farmers. J Soil Water Conserv 66(6):171A–174A

Kassam A, Friedrich T (2011) Conservation agriculture: principles, sustainable land management and ecosystem services. Proceedings of the 40th national convention of the Italian Agronomy Society, Teramo, Italy

Knowler D, Bradshaw B (2007) Farmers' adoption of conservation agriculture: a review and synthesis of recent research. Food Policy 32:25–48

Lal R, Kimble JM (1997) Conservation tillage for carbon sequestration. Nutr Cycl Agroecosyst 49:243–253

Lanjouw P (1999) Rural nonagricultural employment and poverty in Ecuador. Econ Dev Cult Change 48(1):91–122

Larochelle C, Alwang J (2012) The role of risk mitigation in production efficiency: a case study of potato cultivation in the Bolivian Andes. J Agric Econ. doi: 10.1111/j.1477-9552.2012.00367.x

Marongwe L, Kwazira K, Jenrich M, Thierfelder C, Kassam A, Friedrich T (2011) An African success: the case of conservation agriculture in Zimbabwe. Int J Agric Sustain 9(1):153–161

McGinnis L (2006) "Healthy potatoes, healthy people, healthy profit" agricultural research. http://www.ars.usda.gov/is/AR/archive/may06/potatoes0506.htm

Melara W, del Rio L (1994) The use of minimum tillage and leguminous cover crops in Honduras. In: Thurston D, Smith M, Abawi G, Kearl S (eds) Slash/Mulch: how farmers use it and what researchers know about it. CATIE-CIIFAD, Ithaca

GPB, Plan Estratégico de Desarrollo Provincial 2004-2024, Gobierno Provincial de Bolívar, Dirección de Planificación. AH/editorial. Guaranda, Ecuador, 2004

Reardon T, Berdegue J, Escobar G (2001) Rural nonfarm employment and incomes in Latin America: overview and policy implications. World Dev 29(3):395–409

Shaxson F, Kassam A, Friedrich T, Boddey B, Adekunle A (2008) Underpinning conservation agriculture's benefits: the roots of soil health and function. Background document for the FAO Plant Production and Protection Division workshop on investing in sustainable crop intensification: the case for improving soil health, Rome, Italy

Shiferaw BA, Okello J, Reddy RV (2009) Adoption and adaptation of natural resource management innovations in smallholder agriculture: reflections on key lessons and best practices. Environ Dev Sustain 11(3):601–619

Sustainable Agricultural Natural Resource Management—Collaborative Research Support Program (SANREM-CRSP) (2012) Conservation agriculture as a potential pathway to better resource management, higher productivity, and improved socio-economic conditions in the Andean region. Annual technical report 2011–12. SANREM-CRSP, Blacksburg

Swinton SM (2000) More social capital, less erosion: evidence from Peru's Antiplano. Department of Agricultural Economics, Michigan State University, East Lansing

Swinton SM, Quiroz R (2002) Are poor farmers to blame for environmental degradation? Results from the Peruvian Altiplano. Staff Paper 2002–14, East Lansing

West T, Post W (2002) Soil organic carbon sequestration rates by tillage and crop rotation: a global data analysis. Soil Sci Soc Am J 66:1930–1946

Winters P, Crissman CC, Espinosa P (2004) Inducing the adoption of conservation technologies: lessons from the Ecuadorian Andes. Environ Dev Econ 9:695–719

Zimmerer K (1993) Soil erosion and labor shortages in the Andes with special reference to Bolivia: 1953–91: implications for conservation with development. World Dev 21:1659–1675

Enhancing Crop Diversity and Livelihood Security in the Andes Through the Emergence of Agricultural Innovation Systems

Jon Hellin

1 Introduction

The need to increase agricultural production and productivity is taking place at the same time that changes in the global agricultural economy are providing farmers with new challenges and opportunities particularly related to access to markets (Markelova et al. 2009). Enhancing agricultural productivity in marginal areas, such as the Andes, will require new approaches that provide incentives and funding mechanisms that promote the translation of scientific innovations into concrete benefits for poor farmers (Delmer 2005). Agricultural extension, education and training can help many farmers maximize the potential of their productive assets particularly in light of the fact that most new technologies that farmers need will be 'information intensive', i.e. they will require increased levels of knowledge for appropriate management (Tripp 2001). Extension services, therefore, need to support farmers to make better decisions relating to household livelihood strategies (Chapman et al. 2003). However, this need has coincided with deep cuts to publicly funded extension services.

Traditional publicly funded extension provision was criticized for being top–down, paternalistic, inflexible, subject to bureaucratic inefficiencies and, therefore, unable to cope with the dynamic demands of modern agriculture. In contrast, aficionados of private extension provision saw it as the way forward in term of it becoming more demand-driven. The offering of services where demand already exists, however, runs the risk that private sector providers will serve only the better-off farmers and ignore those living in marginal areas (Rivera and Cary 1997). In many cases, the private sector has proven incapable of replacing previous state services due to high transaction costs, dispersed clientele, and low or

J. Hellin (✉)
International Maize and Wheat Improvement Center (CIMMYT), Mexico, D.F., Mexico
e-mail: j.hellin@CGIAR.ORG

S. Mann (ed.), *The Future of Mountain Agriculture*, Springer Geography,
DOI: 10.1007/978-3-642-33584-6_4,

non-existent profits. There is a need for new approaches to extension provision along with consensus on the role of the public and private sectors whereby resource-poor farmers have access to extension provision on a more sustainable basis.

It is possible to separate the provision of funding from the provision of the service (Chapman and Tripp 2003). Debates, therefore, have tended to focus on the differing roles of the private and public sectors in terms of who funds and who delivers extension provision (Umali-Deininiger 2007). This lends itself to pluralistic and diversified extension systems: the separation of delivery and financing of extension results in four main extension modalities:

1. Private delivery and private financing (completely private extension)
2. Private delivery and public financing (contracting out)
3. Public delivery and public financing (typical government extension service)
4. Public delivery and private financing (contracting in).

Practitioners have directed much attention at the extent to which the aforementioned extension modalities facilitate collective action amongst farmers. Collective action is defined as *voluntary action taken by a group to achieve common interests* (Meinzen-Dick and Di Gregorio 2004) and it can play an important role in facilitating farmers' links with public and private sector extension providers. There is a danger, however, that undue focus on the modalities of extension provision and fostering of collective action means that less attention is directed at whether there is a contribution to an increase in social and human capital and rural innovation, such an increase is imperative if farmers are to maximize the potential of their productive assets. There is a need to focus more on the extent to which extension is contributing to farmer empowerment, innovation and change. Discussions are taking place around the concept of agricultural innovation systems (Hellin 2012).

Traditionally extension provision was largely a linear process with knowledge, information, and technology generated from a central source (such as a publicly funded research entity) and then channelled from researchers to farmers via extension agents (Biggs 1988). The problem is that extension agents may make mono-disciplinary theoretical recommendations for what are, in fact, multi-faceted problems embedded in complex local agro-ecosystems and socio-cultural systems. Agricultural development is an immensely complex process characterized by a high degree of nonlinearity. Furthermore, farmers participate in social change as actors rather than as passive subjects i.e. farmers' strategies and interactions shape the outcome of development (Sumberg et al. 2003). Innovation is required so that farmers and service providers can create solutions to production and livelihood problems by taking advantage of new opportunities through the modification of new technologies and existing production systems (Hall et al. 2005).

A linear transfer-of-technology extension approach is likely to stymie local innovation networks (Waters-Bayer et al. 2004). A more effective approach is a systems one that supports the development of a web of dynamic interactions among researchers, extension agents, equipment manufacturers, input suppliers, farmers, traders, and processors (Hall et al. 2005). In a vibrant agricultural innovation system, agricultural development results from efforts to combine technological improvements in production, processing, and distribution with organizational improvements in how information and knowledge are exchanged between various actors in these systems, along with policy changes that create favourable incentives and institutions to promote change (Davis et al. 2008). Agricultural innovation systems, therefore, include both users and producers of information and links them in a dynamic process that depends on learning processes, feedback loops, and iterative interactions that are decidedly non-linear (Spielman et al. 2008).

This chapter explores the future of farming activities in the Andes. Based on qualitative research carried out in the region, it explores new approaches to extension service delivery that stimulate increased agricultural production and marketing, contribute to collective action and which critically also foster the emergence of agricultural innovation systems. The chapter details two contrasting case studies. The first focuses on the role of local farmer extension agents known as *Kamayoq* in generating innovations in agricultural production in the highlands of Peru. The second addresses the feasibility of farmers' in the Andean region maintaining potato crop diversity while also ensuring that they benefit from market opportunities. In this case, an international research organisation played the role of the network broker fostering the emergence of an agricultural innovation system that brings together farmers and the private sector. The viability of farming in the Andes will ultimately depend on an enabling policy environment that encourages public and private sector interventions that lead to the fostering of agricultural innovation systems.

2 The Andean Region

The Andean region falls into three broad ecosystems: the 'Green' Andes stretching from northern Ecuador through Colombia and Venezuela; the 'Yellow' Andes of central Peru and eastern Bolivia; and the 'high climatic risk' Andes of southern Peru and the Bolivian *altiplano*. Altitudes vary from sea level to over 4500 m. Farmers face huge challenges including a seasonal variation in climate that can bring drought, floods, frost or hail within one growing season (Hellin and Higman 2005). The southern altiplano in Bolivia is a particularly harsh area. Annual precipitation ranges from 110 to 250 mm and temperatures fall well below 0° C for 200–250 days per year. Despite these disadvantages, indigenous groups have survived for thousands of years. Much of their livelihood security has been based on the consumption of a range of tubers, principally potatoes, and different grains including quinoa (Dandler and Sage 1985). Farmers have traditionally adapted and selected quinoa and potatoe varieties to reduce their vulnerability to environmental risks.

The common potato (*Solanum tuberosum* subsp. *tuberosum*) is the fourth most important food crop in the world. Potatoes originated in the Andes and have been cultivated in the region for over 8000 years. Within Latin America, seven *Solanum* species are recognised, and with thousands of variants throughout the 4000 km long range of the Andes, they form one of the most diverse crops in the world. In addition to potatoes, farmers have also traditionally grown other tubers such as *oca* (*Oxalis tuberosa*), *papalisa* (*Ullucos tuberosus*) and *isaño* (*Tropaeolum tuberosum*). Quinoa (*Chenopodium quinoa*) is an annual plant that grows in the Andean region between sea level and the heights of the Bolivian altiplano. Quinoa can survive levels of drought, salinity, hail, wind and frost in which other grains would perish (Risi 2001). The grain is used as flour, toasted, or added to soups. Dried, it can be stored for up to 10 years. Quinoa has been cultivated in the Andes for over 7000 years and has long been known and appreciated for its nutritional value (Repo-Carrasco et al. 2001). In the 'high climatic risk' Andes many farmers rely almost entirely on quinoa production, to the exclusion of other crops or livestock.

Farming in the Andes, however, is under increasing pressure. Farmers are growing fewer potato and quinoa varieties and, in some cases, there has been an almost complete displacement of these traditional crops. The reasons for this erosion of diversity include market pressures. As populations of the Andean countries become increasingly urbanised, the urban poor have tended to move away from consuming quinoa and other native grains, and purchase products which are easier to prepare, often cheaper and which are of a more consistent quality though less nutritious, such as pasta and rice. The preference for pasta and other wheat products has extended to rural areas.

Andean highland farmers need access to new productivity-enhancing technologies and to markets. These needs coincide with a substantial reduction in the provision of publicly funded extension services. In Peru, for example, the government agricultural extension program run by what is now the National Institute of Agrarian Research (*Instituto Nacional de Investigación Agraria* (INIA)), employed 1400 extension officers in 1986, but fewer than 100 officers in 1992 (Hellin and Dixon 2008). The state reduced its role to providing only basic services such as certain types of agricultural information (e.g., prices) to support farmers' decision-making (Ortiz 2006). The situation in other Andean countries with respect to publicly funded extension provision is no different but over the last two decades there have emerged new approaches to meet the need and demand for extension provision and the emergence of agricultural innovation systems.

3 The Kamayoq in the Highlands of Peru

Since the 1990s, Practical Action, a development NGO, has been working in Quechua-speaking farming communities in the Peruvian Andes to fill the void in extension provision and to encourage farmer innovation. Initially, the focus was on communities living in the valleys above 3500 m, where common crops are maize, potatoes, and beans. Local families also have one or two head of cattle each, some sheep, and a number of guinea pigs (a food staple in the Andes). Since 2003, the focus has broadened to include communities living above 4000 m, where farmers raise alpacas and cultivate potatoes.

Practical Action recognized that one of the most effective ways to address farmers' needs was through a farmer-to-farmer extension approach that encouraged collective action and farmer innovation. In the early 1990s, Practical Action began training a number of farmer extension agents, known locally as *Kamayoq*. The objective was to train local farmers who would then be responsible for training other villagers and encouraging farmer experimentation and innovation. The training initially focused on irrigation techniques but broadened in the mid-1990s to include priority themes identified by the farmers, including Andean crops, horticulture, livestock, forestry, agro-industry, and marketing.

At the end of 2007, approximately 200 *Kamayoq* had been trained, of whom 20 % were women. Training courses took place at a *Kamayoq* school in Sicuani, 140 km north of Cusco, as well as in different field locations, so that the *Kamayoq* can 'learn by doing' (de la Torre 2004). Instructors include staff from Practical Action, long-serving *Kamayoq*, and experts from regional universities in the cities of Puno and Cusco. During the training, the *Kamayoq* also visit INIA's experimental stations, other NGOs working in the region, and large-scale farmers. Throughout their training, the *Kamayoq* establish contact with technical experts from the private and public sectors and with other farmers. The *Kamayoq* are key players in a two-way flow of information: from the individuals and institutions promoting development, and from the local farmers to these same individuals and organizations. The *Kamayoq* are facilitators of local innovations systems (Hellin and Dixon 2008).

One of the most important impacts of the *Kamayoq* approach has been an increase in social and human capital. Although they provide technical advice, the *Kamayoq* are not promoters of off-the-shelf technologies. On the contrary, they work with groups of local farmers to generate innovations in response to local agricultural and veterinary problems. This is important not just for empowering farmers, but also because farming conditions in the Andes are so complex and diverse that it is difficult to find a ready-to-use technology that large number of farmers can adopt without some degree of adaptation. The *Kamayoq* have encouraged farmer participatory research. Successful initiatives include the treatment of a maize fungus disease; the control of mildew on onions; and the treatment of animal diseases such as *Fasciola hepatica*, commonly known as 'sheep liver fluke', which affects sheep and cattle (and increasingly alpaca).

There is growing demand for the *Kamayoq*'s technical advice along with their skills in facilitating group experimentation and learning. Public and private organizations are increasingly contracting the *Kamayoq* to extend the model beyond the communities and regions where they have operated to date. The *Kamayoq*, like most conventional agricultural extension provision, have worked predominantly on improving and increasing production at the farm level. The next step is, therefore, to determine how the *Kamayoq* model can be broadened to provide farmers with the business services that they need in order to benefit from emerging market opportunities (for example, market linkages and processing skills). What this entails is the further development of the agricultural innovation systems in order to encourage and facilitate farmers' access to markets. The *Kamayoq* could learn much from the Papa Andina initiative in the Andes (Devaux et al. 2009) (see below) which brings researchers together with other agricultural service providers and value chain actors, including small farmers, to promote pro-poor innovations in Andean potato-based food systems.

4 Agricultural Innovation and Potatoes

Recent studies of agricultural innovation highlight the utility of the value chain concept as unit of analysis and focus of interventions aimed at stimulating innovations and developing innovation capacity (Devaux et al. 2009). For example, creating new demand for traditional products through processing and value-adding activities has proved to be an innovative route to higher prices, such as through design of a branding strategy and an awareness campaign for minor millets in India (Gruère et al. 2007). Researchers and development practitioners have extended this concept to the use of native Andean potatoes in the production of potato chips (Devaux et al. 2009).

Andean farmers generally produce four classes of potatoes: mixed native traditional varieties; selected native commercial varieties, improved varieties; and bitter varieties for making freeze-dried potato, known as *chuño* (Brush 1991). They need to find a balance between the environmental risks of growing fewer potato varieties and the risk of not supplying the products desired by the market if they choose to cultivate a diverse range of native potato varieties. Native Andean potatoes can be considered as underutilized because these are important in local production systems but are under-represented in the market (Hellin and Higman 2005).

An innovative approach that development practitioners in the Andean region are exploring is one of 'conservation through use' whereby development practitioners identify market niches for indigenous potato varieties. This has meant that smallholder farmers have been able to benefit from these market opportunities while at the same time maintaining native crop diversity. The Papa Andina project, co-ordinated by the *Centro Internacional de la Papa* (CIP) in Lima, is looking for ways to capitalize on small farmers' knowledge, abilities and the diversity of their potato heritage. A focus of the project is to identify market niches where small farmers actually have a competitive advantage because of their location, local knowledge, access to local varieties or crop management practices. For example, some market niches require small tubers grown at high planting densities and manually harvested, requirements that are much more difficult for mechanised farmers to achieve.

The Papa Andina project facilitates contacts between smallholder potato farmers and the agro-processing companies that are becoming increasingly important buyers of potatoes. Farmers learn more about the processors' demands in terms of preferred potato varieties, volumes required, quality and timing of production. The processors, in turn, learn about the varieties of potatoes that farmers grow and how they grow them. With a greater understanding of the reality faced by both parties, the processors can utilise potato varieties that they have previously ignored.

Through the establishment of an agricultural innovation system, The Papa Andina network has helped participants in potato value chains to develop new market niches for Andean native potatoes grown by poor farmers. T'ikapapa is the first brand of high quality, fresh, native potato sold in major supermarkets in Peru. A national organization called CAPAC-Peru was formed by farmer organizations, NGOs, traders and processors to promote high quality potato products as well as to reduce transaction costs and add value. An agro-processing company, also a member of CAPAC-Peru, contracts farmers to supply potatoes to the supermarket. This company owns the T'ikapapa brand under which the potatoes are marketed. CAPAC-Peru helps organize farmers to supply potatoes meeting market requirements (Hellin et al. 2010).

Tunta is a form of freeze-dried potato produced traditionally from 'bitter potatoes.' Through Papa Andina, farmers' marketing and processing capacities were strengthened, while quality norms for tunta were developed and market studies undertaken. The brand *Tunta Los Aymaras* was developed, which is owned and marketed by the farmers' association '*Consortium Los Aymaras*'. In Bolivia, similar collective action processes were used in market chains for tunta (another freeze-dried potato product) and chuño. A set of Bolivian Quality Standards for chuño and tunta were prepared. Subsequently, chuños were cleaned, selected, bagged, and marketed under the brand Chuñoso. Food quality and safety concerns among consumers have stimulated demand for locally grown organic foods, creating new national market opportunities for indigenous foods. The Andean native potato case studies highlight the importance of a dynamic innovation system that brings together diverse stakeholders in driving technical and institutional innovations.

5 Lessons Learnt from Agricultural Innovation in the Andes

The example of the *Kamayoq* in Peru demonstrate that some sort of private extension delivery is often possible i.e. that information and knowledge are agricultural inputs that by their nature have a market value, with incentives for private participation. The *Kamayoq* in Peru is an example of an attempt to establish and delivering a demand-driven, accountable, and largely sustainable extension system. The *Kamayoq* are private sector actors in terms of their being entrepreneurs who are paid for their services. Farmers pay the *Kamayoq* in cash, in kind, or in the promise of future help through an indigenous system known as 'ayni' and it is this willingness to pay the *Kamayoq* that makes the model largely sustainable. The success of the *Kamayoq* model partly depended on the creation of this market for technical advisory services. This, in turn, ensured a supply of competent advisers (the *Kamayoq*) and the stimulation of a demand for advisory services as farmers become aware of new opportunities and the incentives for their own experimentation and innovation. Farmers' willingness to pay the *Kamayoq* means that their role as both technical advisers and network brokers has become institutionalized in many of the Peruvian farming communities where they evolved.

Within pluralistic extension systems, extension agents need to act as both technical advisers and facilitators/network brokers. The latter role is in many ways the key one as it can contribute to the fostering of collective action and agricultural innovation systems. The *Kamoyoq* and NGO Practical Action acted as network brokers helping establish links between different chain actors in the Peruvian highlands. In the case of the *Kamayoq* the emphasis was on enhancing collective action amongst farmers and creating an agricultural innovation systems. The innovation system, in which the *Kamayoq* play such a key role, has not yet reached the level of maturity of the Papa Andina example but it has the potential to do so. Papa Andina's work, on the other hand, illustrates how collective action involving small farmers, market agents, researchers and other agricultural service providers can generate pro-poor market chain innovations (Devauz et al. 2009).

The Andean case studies illustrate the importance of focusing beyond the modalities of extension provision and analysing the extent to which extension is contributing to the evolution of agricultural innovation systems. While different extension modalities can foster collective action, this in itself is not enough to encourage agricultural innovation; extension needs to focus on combining collective action with networking amongst sets of heterogeneous value chain actors. Both the public and private sectors have a role to play in fostering collective action even though finding a balance between what is publicly funded and what is privately funded is a continuous quest in extension delivery systems (Klerkx and Leeuwis 2009).

Most cases of successful agricultural innovation systems highlight the importance of collective action and the crucial role of a facilitator or network broker who catalyses this collective action, enhances farmers' access to information and technical assistance, and builds the capacity of a group to engage effectively in production and marketing activities. The *Kamayoq* and *Papa Andina* case studies reported in this chapter confirm the importance of the role of a network brokers to generate the establishment of an agricultural innovation system. As the case studies show, this role can be played by a research organisation and/or non-governmental organisation and then, increasingly, by trusted local farmers.

Local villages selected the farmers who eventually became *Kamayoq*. As leaders, the soon-to-be *Kamayoq* already commanded respect within their respective communities. Initially the NGO Practical Action acted as the network broker training the *Kamayoq* and facilitating the establishment of links between the *Kamayoq* and other key actors. Eventually the *Kamayoq* were sufficiently empowered to take on the network broker role. In this sense, Practical Action and the *Kamayoq* are similar to the network brokers described by Klerkx et al. (2009)

in terms of being *systemic intermediaries* that connect different actors involved in the innovation system facilitating the establishment of appropriate linkages and multi-stakeholder interaction in the innovation system.

The *Kamayoq* are not the only example in the Andes of trusted local farmers playing the role of network brokers. Cabero and van Immerzeel (2007) describe the *Pachamama Raymi* capacity building system in Peru whereby local farmer-trainers facilitated and encouraged the emergence of agricultural innovation systems through a strong focus on collective action and joint learning. The emphasis is not whether the facilitators are external or internal agents, nor whether they are supported by the private or public sector. The key is that they need to be catalysts or knowledge brokers rather than instructors, working with communities to achieve the same communities' defined and perceived goals (Anandajayasekeram et al. 2007).

While the literature is clear about the importance of facilitators, the consensus breaks down when it comes to who is best positioned to take on this role. As Klerkx et al. (2009) point out different actors can act take on the role of network or innovation brokers including NGOs and research organizations e.g. the role of CIP in the context of value chain innovations as part of the Papa Andina network (Devaux et al. 2009). Many agree that NGOs with an appropriate skill set may initially be the best facilitators or network brokers. However, both development agencies and the private sector have pivotal roles in facilitating the emergence of innovation systems.

Development agencies are especially important in the early stages but it is critical to engage the private sector in order to reduce the risk that farmers are 'abandoned' when a project ends. The key issue is trust. Devaux et al. (2009) note that in most types of collective action and emerging innovation systems, appropriate leadership is important, but the particular traits of leaders may vary. Leaders should be trusted, able to motivate the members, and have necessary skills for the collective enterprise; this is what characterizes the *Kamayoq* and Papa Andina examples, as such their respective network brokering activities have led to the emergence of successful agricultural innovation systems.

Successful agricultural innovations systems have emerged in other parts of the developing world e.g. around conservation agricultural practices (Erenstein et al. 2008). These share many characteristics with the Andean case studies detailed in this chapter and suggest that agricultural innovation systems have a role to play world-wide. It is interesting to note that the Mexican government has recently launched an ambitious research initiative, *Sustainable Modernization of Traditional Agriculture* (MasAgro) that focuses on developing, improving and spreading innovative sustainable agricultural practices including conservation agriculture and the use of high-yielding maize and wheat germplasm. The aim of MasAgro is to develop a series of agricultural innovation systems. The geographical focus of MasAgro includes hilly and mountainous areas.

6 Conclusions

Farmers in the Andean region face many challenges and the future of farming in this mountainous terrain remains unclear. Research has shown, however, that mountain farmers' agricultural systems can be profitable and that an effective way to contribute to this profitability and sustainability is through the emergence of agricultural innovation systems. Different extension modalities can foster collective action but this in itself is not enough to encourage innovation. Extension needs to focus on combining collective action with networking amongst sets of heterogeneous value chain actors. There is a need for new approaches to extension service delivery that stimulate increased agricultural production and marketing, contribute to collective action and foster the emergence of agricultural innovation systems. In Peru, collective action and the development of an agriculture innovation system required the network broker activities of initially an NGO and then increasingly trusted local farmers known as *Kamayoq*. The region-wide Papa Andina project relied on the network brokering activities of an international agricultural research organization.

There is a danger that by focusing debate on the modalities of extension provision in terms of the role of the public and private sectors, too much attention is given to the funding and execution mechanisms of extension and too little to appropriate extension modalities that contribute to collective action and the evolution of agricultural innovation systems. The agricultural innovations systems perspective, with its emphasis on fostering inclusive networking amongst sets of heterogeneous actors, is proving to be an effective approach to bringing about development changes in agriculturally marginal parts of the world. The viability of farming in the Andes and other mountain region will likely depend on an enabling policy environment that encourages public and private sector interventions that lead to the fostering of agricultural innovation systems.

References

Anandajayasekeram P, Davis KE, Workneh S (2007) Farmer field schools: an alternative to existing extension systems? Experience from Eastern and Southern Africa. J Int Agric Ext Educ 14(1):81–93

Biggs SD (1988) Resource poor farmer participation in research: a synthesis of experiences from nine national agricultural research systems. OFCOR comparative study paper no. 3. ISNAR, The Hague

Brush S (1991) Farmer conservation of new world crops: the case of Andean Potatoes. Diversity 7(1&2):75–79

Cabero J, van Immerzeel W (2007) Building learning networks for small-scale farmers: Pachamama Raymi as an innovative knowledge management system. Knowl Manag Dev J 3(2):52–63

Chapman R, Tripp R (2003) Changing incentives for agricultural extension—a review of provatised extension in practice. Agriculture research & extension network paper 132, Overseas Development Institute, London, UK

Chapman R, Slaymaker T, Young J (2003) Livelihoods approaches to information and communication in support of rural poverty elimination and food security. Overseas Development Institute (ODI), London 69 pp

Dandler J, Sage C (1985) What is happening to Andean potatoes? A view from the Grass-roots. Dev Dialogue 1:125–138

Davis KE, Ekboir J, Spielman DJ (2008) Strengthening agricultural education and training in sub-Saharan Africa from an innovation systems perspective: a case study of Mozambique. J Agric Educ Ext 14(1):35–51

de la Torre C (2004) *Kamayoq*: promotores campesinos de innovaciones tecnológicas. ITDG Latin America, Lima

Delmer DP (2005) Agriculture in the developing world: connecting innovations in plant research to downstream applications. Proc Natl Acad Sci U S A 102(44):15739–15746

Devaux A, Horton D, Velasco C, Thiele G, Lopez G, Bernet, T, Reinoso I, Ordinola, M. (2009) Collective action for market chain innovation in the Andes. Food Policy 34:31–38

Erenstein O, Sayre K, Wall P, Dixon J, Hellin J (2008) Adapting no-tillage agriculture to the conditions of smallholder maize and wheat farmers in the tropics and sub-tropics. In: Goddard T, Zoebisch MA, Gan Y, Ellis W, Watson A, Sombatpanit S (eds) No-till farming systems, pp 253–278. World Association of Soil and Water Conservation special publication no. 3. Bangkok, Thailand

Gruère G, Nagarajanm L, King O (2007) Collective action and marketing of underutilized plant species: the case of minor millets in Kolli Hills, Tamil Nadu, India. CAPRi working paper. International Food Policy Research Institute (IFPRI), Washington, DC

Hall A, Mytelka L, Oyeyinka B (2005) Innovation systems: implications for agricultural policy and practice. ILAC Brief 2. IPGRI, Rome

Hellin J (2012) Agricultural extension, collective action and innovation systems: lessons on network brokering from Peru and Mexico. J Agric Educ Ext 18(2):141–159

Hellin J, Dixon J (2008) Operationalizing participatory research and farmer-to-farmer extension: the *Kamayoq* in Peru. Dev Pract 1818(4&5):627–632. http://www.informaworld.com/smpp/title~content=t713412875~db=all~tab=issueslist~branches=18

Hellin J, Higman S (2005) Crop diversity and livelihood security in the Andes. Dev Pract 15(2):165–174

Hellin J, Higman S, Keleman A (2010) Value chain co-ordination for agro-biodiversity conservation. In: Lockie S, Carpenter D (eds) Agriculture, biodiversity and markets: livelihoods and agroecology in comparative perspective. Earthscan, London, pp 213–228

Klerkx L, Leeuwis C (2009) Shaping collective functions in privatized agricultural knowledge and information systems: the positioning and embedding of a network broker in the Dutch dairy sector'. J Agric Educ Ext 15(1):81–105

Klerkx L, Hall A, Leeuwis C (2009) Strengthening agricultural innovation capacity: are innovation brokers the answer? Int J Agric Resour Gov Ecol 8(5/6):409–438

Markelova H, Meinzen-Dick R, Hellin J, Dohrn S (2009) Collective action for smallholder market access. Food Policy 34:1–7

Meinzen-Dick R, Di Gregorio M (2004) Collective action and property rights for sustainable development. 2020 Vision. Focus 11. Brief 1 of 16. IFPRI, Washington, DC

Ortiz O (2006) Evolution of agricultural extension and information dissemination in Peru: an historical perspective focusing on potato-related pest control. Agric Hum Values 23:477–489

Repo-Carrasco R, Espinoza C, Jacobsen S-E (2001) Valor Nutricional y usos de la Quinua y de la Kañiwa. In: Jacobsen S-E, Mujica A, Portillo Z (eds) Memorias Primer Taller Internacional en Quinua: Recursos Genéticos y Sistemas de Produccción. Proyecto Quinua CIP-DANIDA

Risi J (2001) 'Producción de quinua en el Altiplano sur de Bolivia', in Jacobsen, S-E., Mujica, A., and Portillo, Z. (eds.) *Memorias Primer Taller Internacional en Quinua: Recursos Genéticos y Sistemas de Produccción*, Proyecto Quinua CIP-DANIDA

Rivera WM, Cary JW (1997) Privatizing agricultural extension. In: Swanson BE, Bentz RP, Sofranko AJ (eds) Improving agricultural extension—a reference manual. FAO, Rome, pp 203–211

Spielman DJ, Ekboir J, Davis K, Ochieng CMO (2008) An innovation systems perspective on strengthening agricultural education and training in sub-Saharan Africa. Agric Syst 98:1–9

Sumberg J, Okali C, Reece D (2003) Agricultural research in the face of diversity, local knowledge and the participation imperative: theoretical considerations. Agric Syst 76:739–753

Tripp R (2001) Seed provision and agricultural development: the institutions of rural change. Overseas Development Institute (ODI), London, 182 pp

Umali-Deininger D (2007) Public and private agricultural extension: partners or rivals? World Bank Res Obs 12(2):203–224

Waters-Bayer A, van Veldhuizen L, Wettasinha C, Wongtschowski M (2004) Developing partnerships to promote local innovation. J Agric Educ Ext 10(3):143–150

The Future of Agriculture in the High Atlas Mountains of Morocco: The Need to Integrate Traditional Ecological Knowledge

Bernadette Montanari

1 Introduction

The High Atlas Mountains in Morocco as other mountains throughout the world are landscape providing major economic resources and ecosystem services. Mountain systems represent one-fifth of the world's land and are home to 570 million people. Mountain zones are also important suppliers of water, food, hydroelectricity, timber, other mineral resources and biological diversity, increasingly vulnerable to climate change (Smethurst 2000; Knippertz et al. 2003). As precipitation in Morocco is expected to reduce by 5 % in mountainous areas and 30 % in the southern regions of the country for the period between 2011 and 2050, water as a prime commodity will be greatly affected by these changes. Water shortage is in turn likely to alter agricultural production. Indeed, it is anticipated that agricultural land in Morocco will be reduced to 8 % by the end of the twenty-first century compared to the current 12 %. Further quantitative data suggest a general decrease of water resources from 10 to 15 %, which could in turn affect water availability for irrigation (Knippertz et al. 2003). As a measure to counteract the climatic conditions in the country, the government has launched the Green Moroccan Plan (2012–2020). It proposes a long-term development plan based on agricultural activities to combat poverty and raise the Gross National Product, to create employment, to bring technical support to liberate environment markets, to increase citrus and vegetable exports and to attract major economic capital, even to conserve natural resources (Ministry of Agriculture and Fisheries 2009). The Green Morocco Plan therefore, presents tailor-made solutions and practical

B. Montanari (✉)
School of Anthropology and Conservation, University of Kent, Marlowe Building, Canterbury CT2 7NR, UK
e-mail: bernadettemontanari@hotmail.com

S. Mann (ed.), *The Future of Mountain Agriculture*, Springer Geography,
DOI: 10.1007/978-3-642-33584-6_5,

measures that offer sustainable solutions to the current climatic conditions and food security, but also to promote socio-economic development.

Although it is at present difficult to predict how this will affect the livelihood of mountain communities, it is, however, important to draw attention to how mountain population will cope with such changes. This chapter describes how the inhabitants of a Berber community in the High Atlas Mountains of Morocco have managed to maintain subsistence over the past centuries, for which agricultural activities and traditional ecological knowledge have played a major role for the continuity of the community's livelihood. It further suggests that in the face of climate change and the government's measures to cope with the increasing problem of water scarcity, the communities' traditional way of natural resource and water management need to be integrated into government initiatives. This chapter is partly based on the author's PhD dissertation which has closely examined the mechanisms to implement an essential oil distillation project as part of the government new measures to overcome natural resource erosion, desertification and poverty alleviation. Although the research was conducted in eight villages of the Agoundis valley, this chapter refers only to the traditional activities that occur in El Maghzen, to illustrate the importance of traditional ecological knowledge in the context of climate change.

2 Climate Change and Mountains

Climate change is at the heart of world preoccupation and undeniably the most persistent challenge of the twenty-first century (United Nations world water assessment programme 2010; Neil Agder et al. 2003). Mounting evidence shows that the global hydrological cycle is intensifying and accelerating due to global warming, resulting in increased floods and droughts and dramatic alterations of long-term precipitation patterns throughout the world (Speth and Fink 2010). North Africa is not excluded from these changes and both environmental and anthropogenic systems in the zone are affected. Therefore, a global temperature rise of 2 °C is likely to lead to a corresponding warming of 1–3 °C, and further prediction foresee a temperature increase of 1.4–5.1 % expected to take place in the Mediterranean region by 2055 (Giannakopoulos et al. 2005; Bravo et al. 2008). Morocco, therefore, with a recorded spring rainfall decrease of 40 % and increased dry spell periods (particularly towards the end of the rainy season February to April), since the 1960s, is undeniably affected negatively both at the national and regional levels (Abdelfadel and Driouech 2009). While further predictions anticipate that precipitation in North Africa will further drop by 10–20 % and a rise of temperatures between 2 and 3 °C by 2050, arid conditions will undoubtedly be exacerbated. Agricultural activities, in particular those concerned with non-irrigated rain fed crops are most likely to be negatively impacted by these changes (Schilling et al. 2012).

As it is well recognised, the delicate balance of mountain environment, one of the planet most fragile ecosystems, is affected by any subtle change in climatic conditions (Parish and Funnell 1999). Mountainous regions of Morocco, showing a 5 % rainfall reduction, 30 % in the southern regions, and rising temperatures of 1.2 °C, are ultimately undergoing drier conditions. These warming patterns and fluctuations are more likely to impact the Atlas Mountains than the coastal parts of the Atlantic Ocean and the Mediterranean Sea (Schilling et al. 2012; Knippertz et al. 2003). Subsequently, rural populations, particularly those who depend heavily on natural resources for subsistence, and to a lesser extent those living in lower plains, are most likely to be affected by these changes particularly in agricultural activities, fisheries and other resource domains (Price and Funnell 1999).

What is more, as the country increasingly experiences drier spells, the needs for water are estimated to reach 16.2 billion m^3 by 2020. However, the provision of 17 billion m^3, theoretically available from 2020, would require the construction of further dams and deep wells. While these constructions are thought to meet up to 60 % of the drinkable water for the major urban areas, and supply for irrigation of large agriculture settings and power generation for the national power grid (Jellali 1999; Nusser 2003), it does not necessarily imply that water is equally distributed, and in fact, it may increase deficiencies and inadequate performance, impact on ecological systems, exacerbate natural resource degradation, displace local inhabitants, and disrupt the social, cultural, and economic life of communities living within the proximity of these structures (Minoia and Brusarosco 2006; Tilt et al. 2009).

3 The High Atlas Mountains: Geographic Situation

Morocco possesses the largest mountain area in North Africa, broadly divisible into three parts with the following characteristics:

A. The Middle Atlas, which rises to 3,000 m with an annual precipitation of 600–1000 mm per year on the west side but decreasing to 300–500 mm a year in the east. These mountains form a major barrier between Mediterranean and Atlantic Morocco and the Sahara (Fig. 1), and are the location of some of the highest North African peaks, ranging from 700 m above the permanent snowline at 3,300 m. Season, altitude and rain precipitation largely determine the climate and can vary significantly in some valleys.
B. The Anti-Atlas, a plateau characterised by dissection, and situated south of the High Atlas, with an altitude ranging between 500 and 1,500 m, and displaying a complex climate, with precipitation of 120–650 mm per year. The Anti-Atlas landscapes play a fundamental role in the Moroccan environmental system, protecting the country from dry, hot Saharan air, and collecting rain, which in turn feeds most of the streams, and main land water, which lowland and agriculture depend upon. Further, the Anti-Atlas provides much of the food grain, fibre and meat consumed in the country.

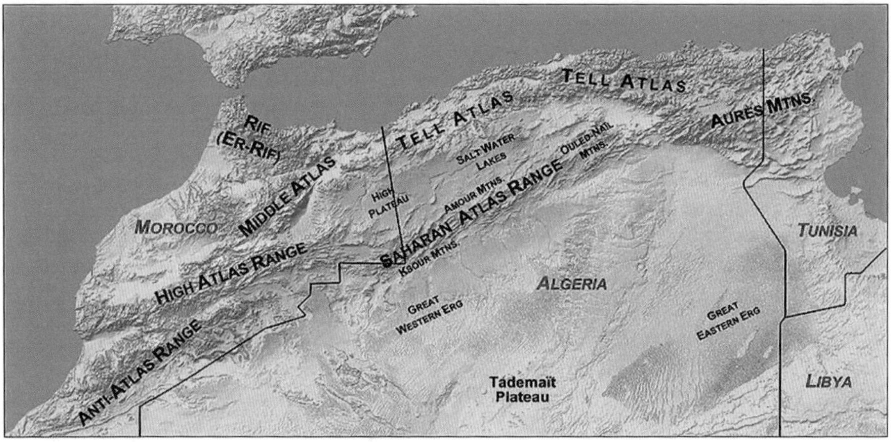

Fig. 1 The range of the Atlas Mountains in Morocco. *Source* http://en.academic.ru/dic.nsf/enwiki/762566

C. The High Atlas rises in the West towards the Atlantic Ocean and extends in an eastern direction as far as the Moroccan–Algerian border. An abrupt drop from the Atlantic to the southwest marks an impressive transition right up to the coast and the Anti-Atlas range. The Jbel Toubkal at 4,167 m is the highest mountain included in this range and hosts the Toubkal National Park. The High Atlas is an important barrier in the Moroccan weather system, preventing the pronounced Saharan conditions, particularly in the summer from influencing the Mediterranean climate to the north. This results in dramatic changes in temperature across the range. Snow falls regularly in the highest elevations of the range, which permits winter sports. Snow can last until late spring and is present mostly on the northern faces of the range. In this respect, the High Atlas represents the backbone of Morocco. An estimated 800,000 ha of irrigated land are in the mountain regions, which can in turn support 30 % of the population. In the High Atlas, traditional irrigated terraced agriculture can support up to 28 persons per square kilometres (Downs 2003; Barrow and Hicham 2000).

4 Mountains: A Complex Socio-Political Environment

Mountain environments present unique political, economic and cultural features. Geographically, they often form ‘natural’ borders and frontiers between territories, and define political divisions between states and regions and between political units within a state. While geographical access, slow and precarious systems of transportation remain problematic, physical and political remoteness from central power and authority contribute to these communities’ disproportionate political representation. Culturally, they stand apart from the lowlands, physical environment accentuating the boundaries between cultural groups. Mountain environments are difficult and complex areas to live in and often perceived as marginal by outsiders. These marginal social groups have limited options, which further encourage them to depend on local resources, leading to landscape degradation, and limited accessibility and isolation increases the community's crucial dependence on natural resources (Jodha 2007; Collins 2008). Marginality, therefore, both in a physical and socio-political sense, exacerbates poverty in these difficult zones and accentuates the importance of cohesion for collective self-help and reliance.

The High Atlas Mountains in Morocco serve not only as a border between Morocco and Algeria but also to separate traditional Berber communities from the rest of Moroccan society. Indeed, they represent a very important refuge for these communities. These mountain areas which offer significant untapped human and natural development potential abound in natural resources, particularly aromatic and medicinal plants. The people and the landscapes have always been changing, shifting, and self-regenerating. Because these regions face strong environmental constraints, the populations are called to manage their environment, something that they do generally well, contrary to the widespread current belief that local people mismanage their resources (Ostrom 1990; Scoones 1994; Pretty and Pimbert 1995; Pretty and Shab 1997; Ghimire and Pimbert 1997).

5 El Maghzen in the Agoundis Valley, High Atlas Mountains

El Maghzen is a Berber village in the Agoundis valley, about 100 km from Marrakech. It is part of the Talat n'Yakoub circle[1] in Al Haouz province and is close to the Toubkal National Park, a biodiversity 'hot-spot'. The disparity between the cities and these areas is considerable, mostly because the populations are poor, marginalised, and lack the basic infrastructures for development. The inhabitants represent one of the poorest segments of Moroccan society in terms of literacy, infant mortality, availability of potable water, electrification and other development indicators (Russell 2004). Although the environment is biologically rich, especially in aromatic and medicinal plants, the region's natural resources overall are declining owing to over-harvesting in the face of the increasing demand for phyto-aromatic products and the needs of a growing population. Local people harvest plants during the summer months, for both herbal medicine and for trade, the most important being thyme (*Thymus satureioides*), although sage (*Salvia aucheri*) and a species of lavender (*Lavandula dentata*) are also collected. These

[1] A circle or 'cercle' in French was the smallest administrative unit of the French colonies in Africa. A circle was usually composed of districts, and these composed of several villages (Personal communication, Alifriqui 2008).

plants are one of the few sources of cash income. The plants are traded down the valley via several middlemen to urban markets in Marrakech and beyond. The trade follows two commodity chains, one official and the other informal and illegal (Montanari 2004). Although this income varies in terms of the amount of plant material collected, it nevertheless represents an important contribution to the household economy.

In the past, the inhabitants of the villages in the Agoundis valley had collective rights of access to the land for their subsistence needs, e.g. harvesting medicinal plants and collecting wood for fuel and building material. An autonomous system of Berber customary law, known as *jama'a*, regulated rights of access to land for grazing, forests and water. This traditional law, never codified, was rather flexibly applied for solving problems of resource use. It regulated not only individual access but also collective access, and was integrated into the cultural and political life of the community (Id Balkassm 2002). In 1917, during the French colonial period, the national government claimed to own the land and the Department of Water and Forestry was assigned the task of control. Since then, village residents have had limited access to their traditional lands, and only for the collection of dead wood and medicinal plants. In times of conflict and confusion over land access, during the Protectorate and at the present time, people have fallen back on customary law to access the resources. Currently, local people can only collect plants for personal use. If caught collecting for trade, they are fined by the Department's representative.

6 The Socio-Economic Characteristics in the Agoundis Valley

The High Atlas communities demonstrate many of the 'poverty and livelihood' characteristics found among mountain communities. Households living under the poverty line represent 31.6 % of the total and the number of households considered vulnerable represents 25.4 % of all inhabitants. In these terms, the Agoundis valley is one of the poorest segments of the Moroccan areas and this is reflected in literacy levels, infant mortality, availability of potable water and other development indicators (Russell 2004). With its relative inaccessibility, small population, and a subsistence economy based on mixed farming and local natural resources, villagers manage to sustain themselves, focussing on barley, almonds and walnuts in terraces of carefully irrigated fields. Forests provide wood for cooking, construction, and heating, as well as forage for animals. Water management and irrigation is both very fragile and complex (Crawford 2003; Saxena et al. 2001). During the summer season, when temperatures can reach over 45 °C and the river is dry, water is scarce, especially if rainfall has been low during the winter months. In this case, water has to be collected from the river in containers. To provide for cattle, women have to spend many hours daily collecting fodder. The winter season can be difficult as there is no food for the animals. Women will resort to gathering fallen walnut leaves. For baking the daily bread (*tanourt, arum*), wood collection is essential and requires long hours spent high up in the mountains, not without any risk of accident, as people sometimes have to climb up mountainsides to cut the wood. Women predominate in agriculture and in collecting vital resources such as wood and fodder.

Household cash income is low, partly provided by outside members of the family who are in employment and live in the cities or further away. Resources such as almonds, walnuts, carob pods and aromatic plants are traded to generate income. Individual households and their family connections will often break almonds and walnuts together so that the head of the household can take the merchandise to sell at the local *souk* (market). Given that 50.63 % of men and 14.19 % of women are active in the labour market, the rate of unemployment at 6 % is low. The population is mainly young and very active. Twenty-three percent of the population has received schooling. This rate is lower for women; with only 6 % having access to education compared to 44 % for men (Benaboubou 2004). Although there are adult literacy programmes, administered by a woman from the village, illiteracy is high: 84.39 % amongst women and 50.67 % amongst men (HCP 2007). Because the local mother-tongue, Tachelhit, is not officially recognized, the lessons mainly focus on the Koran and basic arithmetic and Arabic. In the whole valley, most inhabitants speak Tachelhit and a small part of the population speaks dialectal Arabic.

While the central government has planned for the remote parts of the country to have electricity and running water by 2012, only 24 and 37 % of respectively households benefit from these commodities in the Al Haouz region. In the Agoundis valley, the main sources of light remain candles (36.5 %), gas (80.8 %) and solar panels (18.3 %). Spring water is the main source of water supply for 80.3 % of the population and village fountains for 23.8 % (HCP 2007). The Agoundis valley and its villages do not have medical facilities. From El Maghzen the nearest such facilities are 8 km away in Ijoukak and 13 km in Talat n' Yacoub.

Nowadays, families have no more than three to four children, but in the past, it was common for women to have up to ten or twelve. Some women would sometimes lose as many as five. Women still give birth at home and child mortality remains high, at 7.5 % (HCP 2007). Children are central to production in this rural environment and parents view their children as a source of wealth and power. Moreover, children cost less in the rural areas than in towns. Children may go to the local school to learn Moroccan Arabic, mathematics, and to read and write. The local commune encourages parents to send their children to school. However, very few children can make it to college for further education. From an early age, children work in agriculture, wood and water collection. The parents do not have the money to send their children to further education. Children are regarded as being more useful for their contribution to the household. Girls take care of their younger siblings, thereby relieving a mother pregnant with another child; or they may work in the fields. Boys tend to labour in the gardens or at the river. Members of Berber families work together but also divide the rewards together and the elderly rely on their adult offspring for care.

7 Traditional Ecological Knowledge and the Management of Activities

Traditional or local ecological knowledge is a complex dynamic learning process embedded within cultural forms and social institutions (Berkes et al. 2000; Davis and Wagner 2003; Ellen and Harris 2000; Ellen 2011). In many places it is vital for the maintenance of the land, water and biological resources upon which people depend, and enters into decision-making chains at every stage in production, management, distribution and consumption (Grenier 1998; Turner and Garibaldi 2004; Folkes 2004).

In El Maghzen, as in most other traditional societies, the transmission of knowledge is both *horizontal* between the members of the same generation, and *vertical* between members of different generations, stereotypically between parents and their offspring (Guglielmino et al. 1995). Both these horizontal and vertical transmission pathways contribute to the reproduction of traditional knowledge, which—though adaptable—is inherently conservative (Guglielmino et al. 1995; Eyssartier et al. 2008). This knowledge and transmission extend to all activities.

7.1 Traditional Agriculture: A Balanced Management System

Owing to its topographic and geographic position, Agoundis is one of the narrowest and enclaved valleys of the High Atlas, enclosed between abrupt forested slopes, and offering very little cultivable space. The duality of this spatial structure produces noticeable differences in the landscape and in the availability of resources. However, these harsh and fragile environments are more often than not heavily settled. The strong declivity of the slopes favours the streaming and erosion of the ground, thus necessitating the construction of terraces. Because of the altitude ranges, local families have traditionally diversified livelihood strategies according to the seasons. With the integration of terrace agriculture in the landscape, local populations have managed to subsist on diversified rotation agricultures (Barrow and Hicham 2000). Millennia of human modification have shaped the typicality and diversity of these landscapes to control erosion and to promote agriculture. The Agoundis valley has, therefore, access to a remarkable anthropic landscape (Gerbati 2004). Farming takes place in terraced fields cut into the steep valley sides (Plate 1). For centuries, the villagers of the High Atlas have practised a mixture of subsistence cultivation and pastoralism.

Although the inhabitants of El Maghzen tend to rely heavily on local resources, they have always been self-sufficient and techniques for managing the land and other natural resources have little altered over the centuries. They have developed flexible mechanisms—terracing, irrigation, pasture management, transhumance, crop selection—that have allowed them to survive difficult (including extreme climatic) conditions using the resources at hand; villagers have adapted and shaped

Plate 1 A typical landscape of wheat terraces in the Agoundis valley. *Source* Montanari© 2007

the landscape according to their needs. Gardens, not only provide the basic crops to feed the family, but also a space for recreational and other social activity, and where, for example, women can make pre-arrangements for meetings. In the spring and summer, people will spend a considerable amount of time in the garden and have food and tea brought to them by a younger member of the family. Every household has a more or less equal number of terraces and due to the restricted space for cultivation, the gardens are well-delineated, ploughed with a mule and wooden plough. Nothing is added to the land apart from cow and chicken manure once a year.

The main crops in the gardens are wheat and barley, harvested once a year during the summer, and more recently alfalfa (*Medicago sativa*), which is collected mainly for cow fodder. In the absence of chemical pesticides, crops are often companion-planted in groups of two or three cultigens, for example tomatoes with maize and peas, broad beans, onions and egg plants. While most women are involved in garden activities such as sowing, weeding and harvesting, it is men who have the duty of ploughing. This is performed with a donkey or a mule and traditional plough (Plate 2). Harvesting wheat and barley may be undertaken both by men and women. This activity is performed during the summer (May, June) and involves cutting the wheat and barley to be carried back to the village terraces for drying. Women will often gather and go down to the terrace together, and may reap either alone or with other family members. Although men may also harvest wheat and barley from the terraces, the wheat and barley processing done by men is quite different to that undertaken by women, as they are more involved in threshing. The work is mainly undertaken using donkeys but does require control

Plate 2 Ploughing with donkey on terraces in El Maghzen. *Source* Montanari© 2008

of the animal and a certain amount of strength to push the axle of the mill. The collection of wood is either a female or male activity, in which a group of young children team up for the task, loading the donkeys and mules with chopped wood before returning to the village at dusk.

Although female outdoor activities reflect the harshness of daily existence, many activities are collective and group cohesion is very important in the community. For instance, cow fodder collection also occurs in the mountains and groups of between two and four women may collect together. It requires traipsing long distances over the mountain slopes and carrying the fodder back to the village. Cattle fodder collection also occurs from the gardens, and while these are closer to the village, it still requires carrying bales of fodder on the back. The feeding of animals is undertaken within the village which is close to home. It usually takes place early in the morning or later in the afternoon (Plate 3).

7.2 Other Traditional Edible Crops

In addition to the main field crops, there is a range of non-field edible resources, which are regularly consumed by the community and often exchanged and sold for cash, making an important contribution to people's livelihood.

Plate 3 Cattle fodder collected from the terraces in El Maghzen. *Source* Montanari© 2008

7.2.1 Almonds and Walnuts

Almonds (*Prunus amydalus* var *dulcis*), *luz,* and walnuts (*Juglans regia*), *tarkayin* in Tachelhit, are abundant in El Maghzen. Walnut and almond trees usually belong to the same family, and are harvested in autumn. The cracking of almonds and walnuts is often a collective task and a family event in which all members participate, sometimes helped by outsiders. The almonds and walnuts are therefore cracked on a stone, an occasion to converse about village or family matters over tea. The nuts are then sold to the local *souk*, for about 60 Dh a kg. They represent an important source of income for families, particularly in times of hardship.

7.2.2 Barbary Fig Tree

The Barbary fig tree (*Opuntia megacantha),* or *aknari* in Tachelhit, has its origins in the Canary Islands but is found throughout Morocco. It is very common around the village and widely consumed by the community. It is collected mostly by men. In the village, the collection of the fruit (Plate 4) with a wooden V-shaped stick is a very delicate process as the fruits are covered with prickles. Once these have been removed, the fruit can be consumed when ripe. They can be purchased in Marrakech, sold loose on carts pulled by donkeys or mules. The fig has medicinal value in the treatment of diarrhoea although will produce the opposite effect when consumed excessively.

Plate 4 A special hand made tool for handling the prickly Barbary fig. *Source* Montanari© 2007

7.2.3 Capers

Capers (*Capparis spinosa*), or *tylilout* are widespread around the village. The young flower buds are collected in early summer and packed into plastic bottles in a mixture of salt and water that is regularly drained and changed. They are then used for culinary purposes and make an pleasant addition to omelettes and *tajine*. They have beneficial properties and are particularly used as an anti-rheumatic.

7.2.4 Carob

Carob (*Cerotonia siliqua*), or *tikidit* (*tikida* for the pods), are already much exploited. They are collected in the village once a year, and sold at 25 Dh a kg to local middlemen, who transport the merchandise to bigger collection points from where they are then exported to Europe.

7.2.5 Figs

The common fig tree *(Ficus carica), ukzern* in Tachelhit, belongs to the Moraceae. Trees grow up to 6 m, flower between June and September and the seeds ripen in August and September. The succulent fruits are actually not seeds or flowers at all, but a receptacle which encloses a multitude of flowers and seeds which never see the light but still ripen perfectly. It has great nutritional value, is highly calorific and is used as a laxative and expectorant. In El Maghzen, where fig trees are abundant, fruit is collected and left in the sun to dry and consumed in a mixture of almonds and walnuts served with tea.

7.2.6 Olives

Olives from the species *Olea europea*, with its vernacular name *zaytun*, are also harvested in December and January. It is often undertaken by women and men from the same lineage. Olive oil is then produced in the village oil mill in the traditional manner, each family taking turns pressing the olives. The first operation involved in pressing the olives is conducted using a donkey. The donkey is led repeatedly around the mill to squash the olives in the millstone. Thyme (*azoukni)*, and lemon are then added for flavour. Once the first press has been finished, the broken olives are placed into baskets which are piled up together. In El Maghzen, a building has been constructed around a fallen carob tree being used as the press. The olive oil so produced is usually for personal use.

7.2.7 Pomegranate

Pomegranate trees *(Punica granatum),* or *taroumant,* which belongs to the Puniceae, are found in the village of El Maghzen and people consume the fruit. From an ethnobotanical perspective, it is the remedy par excellence for ulcers, gastrointestinal ailments and diarrhoea (Belhkadar 1996). Pomegranate juice has become very popular and consumed as a juice.

7.3 Traditional Water Management

The water ditribution shows the importance of traditional water management practices. There are different types of collective water resources. These consist primarily of the Agoundis and Ait Ahmed *oued* (streams). A hydraulic unit for collective management is defined by a central *targa* (irrigation canal), fed by a water pump coming from an artificial pool or from a water catchment associated with a more recently-built cement dam. These initiatives have been funded by the Fonds International de développement de l'Agriculture (FIDA) and the Agence de développement solidaire (ADS) and replace the old traditional system. However, there are two types of water distribution system. The first is called *tawala* and supplies water on request when it is abundant, particularly during the springtime. The second is called *nouba* in which distribution is based on water rights and timed allocation according to lineage. It is the most common system. The allocated time varies from half to a whole day, depending on the accounting unit from the artificial pool. Water originating from villages higher up the valley is diverted from the river to irrigate the terraces of lower villages through an intricate branching system. In the summer when the river is dry, most men pursue the building of river dams. It involves digging the river bed to find water before diverting it to a common pool. This involves lifting huge rocks from the river bed to facilitate terrace irrigation. The architecture of the terraces is such that irrigated water can reach every garden and in each village, the sluices are opened in turn at precise times of the week so that everyone gets a share. Men tend to undertake heavier and more complex structural tasks, such as maintaining the main elements of the irrigation system which may involve some digging. With current fears regarding climate change, there are concerns not only in terms of irrigation but also of drinkable water for the village. As shortage of water is an issue during the summer months, maintenance of the waterworks is vital to ensure that the river flow can reach the lower terraces.

7.4 The Agdal

In the High Atlas, the *agdal* designates a method of appropriation and management of the land, a status resulting from customary rights. Although nomadic seasonal transhumance was very common up to the 1950s between natural low and highland pastures, herding is now mainly sedentary, involving small flock of grazing goats or sheep. It takes place during particular periods, especially in the higher pastures of the valley, and is often supplemented by the addition of fodder harvested from the garden or the mountains, or even with hay when villagers can afford it (Bourbouze 1999). In part of the High Atlas where transhumance still occurs, another type of resource management institution known as the *agdal* system functions at a larger territorial level that of the fraction and tribe (Dominguez et al. 2012; Genin and Simenel 2011). This method of managing common resources is implemented by the *jama'a* which regulates access to a lineage territory and its resources. It is a geographical and agro-ecological space characterised by the physical environment and specific biotic resources (trees, pasture, and agriculture). Rigid opening and closing dates for usage of specific-collective pastures regulates and sustains these grazing practices. *Agdal* are typically found in high mountain pastures and are the most widespread and formalised system of transhumance where good pastureland and water can be found after the winter snow and during the dry summer months. This system is important because traditionally Berber pastoralists followed a pattern of seasonal migration, grazing herds at low altitudes during the winter and at higher altitudes in summer, allowing for the regeneration of pastureland during the months when the *agdal* were closed (Mahdi 1999; Auclair 1996).

Plate 5 Azoukni (*Thymus satureioides*). *Source* Montanari© 2008

7.5 The Collection of Thyme and its Economic Contribution

In El Maghzen, local people collect aromatic plants which represent a considerable income for the households. Although Thyme, (*azoukni*), does not quite fall into the category of an ecological or cultural key stone species (Paine 1995; Turner and Garibaldi 2004), nor is it employed in religious ritual in El Maghzen, it has a high social profile because of its wide use in the community, particularly as a medicine and as an important source of cash income (Christancho and Vinning 2004).

Thyme (*Thymus satureioides),* typically referred to as *azoukni* in Tachelhit (Plate 5) is collected during the harvest season (mid May to mid July). This vernacular name is used for the male plant characterised by its purple flowers, as opposed to the female plant that is called *tazouknit* (*Thymus pallidus*) which has white flowers. Although collectors are paid on average only 1 Dh/kg, the thyme harvest is important for both women and men as there is no other comparable income generating activities in the valley. It is a valuable contribution to the household economy that ranges from 5481 Dh[2] for men and 4352 Dh for women over the 2 months period. Intermediaries with trucks usually collect in the most remote villages to sell to other intermediaries at the local *souk* and then transport the product to bigger towns such as Marrakech or Casablanca, especially for essential oil distillation (Montanari 2004, 2012).

---

[2] 1 Dirham (Dh, MAD) is equal to £ 0.73 and € 0.090 (May 2012).

Table 1 Women's and men's agreement responses to questions on thyme harvesting methods in El Maghzen

	Women (%)	Men (%)
If entire plant is removed there will be no harvest the following year	24	38
If the tops are cut this will ensure that the plant grows back the next year	73	59
Removal of the entire plant will maximize my income	3	3
Removal of the entire plant is forbidden	0	0
If entire plant is removed middlemen will not want to buy	0	0
If Water and Forestry see that entire plants have been harvested they will fine us	0	0

From an ethnobotanical perspective, *Azoukni* is taken regularly, that is more or less on a daily basis, as a fresh herbal tea infusion or in the dried herb form. In the absence of other form of conventional treatment, the dried herb is powdered and taken regularly by women for painful menses, to relieve gastric disorders (stomach ache, bile complaints, indigestion, intestinal trouble), and respiratory disorders such as colds, coughs, chills and headache because of its warming character.

7.6 Thyme Sustainability

While women tend to harvest on the mountains within the vicinity of the village, men tend to collect bigger quantities on the higher mountain summits. The villagers, however, showed a strong concern about climatic conditions affecting thyme availability because they tend to collect the bulk of the harvest at higher altitudes.

A study conducted by Khadouri (2007) shows that the harvest needs to take place at strategic points on the mountains, and particularly at higher altitudes (2000 m), where the plants can reach up to 60 cm in height because of the climatic conditions affecting its regenerative abilities. In El Maghzen, the harvesters were well aware of the harvesting-extraction procedure and the ensuing problem of thyme sustainability. Most people stated that they did not extract whole plants, and that they were only cutting the tops and that roots should be left in the ground to allow the plant's regeneration for the following year. A handful of adults in El Maghzen admitted that they removed the whole plant to maximise production. The harvesting thyme being economically important for the villagers, there is a genuine concern about the plant sustainability and harvest occurs in a way to maintain availability rather than succumbing to pressure from the forestry authorities (Table 1).

8 Plant Knowledge as an Indicator of Traditional Ecological Knowledge Transmission

Both women and men in El Maghzen share a common knowledge of plants and the distribution of this knowledge follows different patterns. Children, for instance, who usually accompany their parents to the terraced gardens and take part in the various activities there, absorb plant and other knowledge while weeding and preparing the terraces, or when gathering cow fodder or wood with their parents on the nearby mountains. These are important locations also for practical knowledge transmission.

The transmission of plant knowledge through women is entirely within the community. Within families, parents are a strong vector and girls learn particularly from their mother (41 %). However, plant knowledge is also acquired from the grandparents (11 %) and to a great extent, girls learn by themselves (19 %). They acquire this either by watching other women preparing medicine in the house and when collecting medicinal plants in the garden, at the river or in the mountains. There seems to be a combination of these transmission pathways, though the data shed an interesting light on people's own assumptions about how they acquire knowledge. Overall, the pattern shows the importance of the family group. Other means of learning for women are through friends, from the elderly people of the village, and for older women from the herbalist (8 %). Men's knowledge on the other hand, is acquired from their mother but to a lesser extent (26 %). They also acquire knowledge from their grand-parents (6 %). As with women, transmission occurs mainly within the family, defined as parents, sisters, paternal aunt and uncle and the transmission from these combined is greater than with women (38 %). Many men claimed to learn by themselves (12 %), but also from friends and acquaintances in the village (18 %), and to a lesser extent from their fathers.

Men are more likely to have acquired their knowledge while harvesting on the mountainside compared with women (60 % for men and 23 % for women), while women are more likely to have acquired their knowledge in the home than men (42 % for women and 12 % for men). However, knowledge is also acquired in other contexts (35 % for women and 28 % for men), such as the gardens and at the river, although gardens are more important for women than for men. On the other hand, men claim to acquire more knowledge and skills in the setting of the river, and to a lesser extent in the gardens, than women, as they tend to concentrate their activities in these locations (Fig. 2). Some transmission occurs also in the village and this tends to affect women more than men. This is explained by the fact that women tend to spend more time in the village, despite visits to the mountains, river or garden.

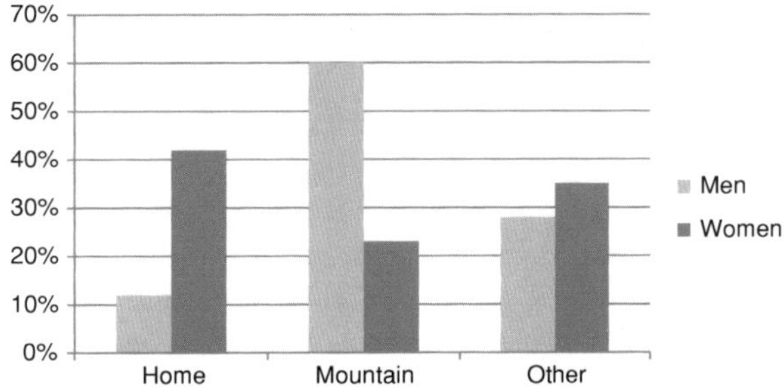

Fig. 2 Response by men and women to the question 'Where did you learn about plants?'

9 Discussion

This chapter has demonstrated the importance of traditional practices in the village for economic well-being, how knowledge about them is distributed and how these practices are valued. In the Agoundis valley, the management of the natural resources is a crucial component of local traditional knowledge that had sustained subsistence for centuries. The transmission of plant knowledge for instance, is intricately wowen into the fabric of the community's traditional ecological knowledge. For the communities these forms of management are part of their cultural heritage, and attachment to the land and the services they render are strong. The inhabitants of the Agoundis valley have always been self-sufficient. They have managed agricultural, water and pasture resources, not only in a way that addresses appropriately the problems associated with living at high altitudes and limited space, but in a way that also buffers extreme climatic conditions through traditional customary law, the *jama'a.* As Parish and Funnell (1999) pointed out, one could anticipate that in the face of climate change, the community has already altered its management practices to accommodate climatic changes. Indeed, local knowledge, skills and capacities have been constantly shaped and reshaped, to respond to particular local problems and situations, in order to achieve and maintain specific survival objectives (Sillitoe 2004).

The villagers for instance, were aware of their dependence on the resources for their subsistence. As an economic practice, thyme harvesting relies on the same traditional skills traditionally sustained the resource. Thyme, as with other resources, is essential for the communities to ensure cash income. The traditional skills applied to the harvest of thyme have been modified to ensure the continuity of the resource on which economic subsistence depends. The way the villagers harvest, is not systemically unsustainable. For the harvesters, a main indication that a change is occurring and likely to affect the availability of thyme is reflected in the lack of rain. They have witnessed this increasingly in the shortage of water and the associated problems for irrigating gardens and this confirms their suspicion that it may also be affecting thyme availability. This observation is based on a locality shaped by past and present experiences, and interpreted in relation to the physical and social environment. For the inhabitants of the Agoundis valley, the threat to thyme sustainability is perceived through changes in weather patterns that they cannot measure in technical terms. The communities, however, lack the technical skills to enhance their long-term sustainability and conservation, particularly in relation to a marginal natural environment.

As the country is adopting the Green Morocco Plan (2010–2020) to counteract the problems of climate change and water scarcity, there is an urgent need to accommodate traditional practices with new development initiatives. Such improvements may be achieved through interdisciplinary work, a hybridisation of knowledge, where technical and scientific skills are combined with traditional knowledge (Sillitoe 2004, 2009; Sillitoe and Bicker 2004; Sillitoe and Marzano 2009).

9.1 Forecast

Given the predicted climate change described above, thyme sustainability is likely to become increasingly unsustainable and there is an urgent need to apply appropriate solutions to conserve the species in El Maghzen. While cultivation seems an obvious solution, it does not stand as a viable option at village level, mainly due to land availability being inadequate. Although thyme regeneration is more profuse on higher mountain summits than at lower levels, the topographic environment remains difficult and cultivating plants at such altitudes remains problematic. It may even jeopardise women's harvesting, given that they tend to collect more locally. What is more, it is difficult at present to predict if land access and thyme harvesting will be further restricted by the Department of Water and Forestry Authorities. However, measures to do so will not prevent the population from harvesting thyme because of its important economic contribution to the household. The Green Moroccan Plan to implement massive agricultural production could therefore prove to be challenging in the mountainous regions of the country. As globalisation is gradually gaining ground, even in the more remote mountain areas, people are becoming increasingly exposed to new commodities and new technologies. Whilst the traditional agriculture could benefit from some technological input, a major threat to the conservative agricultural management is the overuse of pesticides and other chemical growth enhancers. This may further encourage a strong reliance and dependence on these substances, jeopardising not only the traditional agricultural management but also endangering the traditional Berber way of life of the Agoundis valley.

References

Auclair L (1996) L'appropriation communautaire des forets dans le Haut Atlas marocain. Cahier des Sciences Humaines 32:177–194

Barrow CJ, Hicham H (2000) Two complimentary and integrated land uses of the Western High Atlas Mountains, Morocco: the potential for sustainable rural livelihoods. Appl Geogr 20:369–394

Bellakhdar J (1996) La Pharmacopée Marocaine Traditionnelle. Ibiss Press, Casablanca

Ben Abdelfadel A, Driouech H (2009) Climate change and its impacts on water resources in the Maghreb region. Communication presented at the 5th world water forum, Istanbul, Turkey, 16–22 March 2009. http://www.worldwaterforum5.org. Accessed May 2012

Benaboubou M (2004) Etude ethnobotanique du Haut Atlas. Diplôme des études supérieures spécialisées. Gestion de l'environnement et développement durable. Faculté des lettres et sciences humaines université Mohamed V, Agdal Rabat

Berkes F, Colding J, Folke C (2000) Rediscovery of traditional ecological knowledge as adaptive management. Ecol Appl 10(5):1251–1262

Bourbouze A (1999) Gestion de la mobilité et résistance des organisations pastorales des éleveurs du Haut Atlas marocain face aux transformations du contexte pastoral maghrébin. In: M. Niamir-Fuller (ed) Managing Mobility in African Rangelands. Intermediate Technology, London, pp 18–46

Bravo DN, Miguel B, Teodoro Lasanta A, López Moreno JI (2008) Climate change in mediterranean mountains during the 21st century. J Hum Environ 37(4):280–285

Christancho S, Vinning J (2004) Culturally defined keystone species. Hum Ecol Rev 11(2):153–164

Collins TW (2008) The political ecology of hazard vulnerability: marginalization, facilitation and the production of different risk to urban wildfires in Arizona's White Mountains. J Political Ecol 15:22–43

Crawford D (2003) Arranging the bones: culture, time and inequality in Berber labour organisation. Ethnos 68:463–486

Davis A, Wagner JR (2003) Who *k*nows? On the importance of identifying 'experts' when researching local ecological knowledge. Hum Ecol 31(3):463–489

Dominguez P, Bourbouze A, Demay S, Genin D, Kosoy N (2012) Culturally mediated provision of ecosystem services: the 'agdal of Yagour'. Environ Values J 21:277–296

Downs CM (2003) Hearth fuel acquisition and use in Morocco's Immane valley. Master of Science in Forestry Dissertation, Michigan Technological University

Ellen R (2011) 'Indigenous knowledge' and the understanding of cultural cognition: contribution of studies of environmental knowledge systems. In: Kronenfeld DB, Benardo G, Fisher UC (eds) A companion to cognitive anthropology. Wiley, Oxford, pp 290–313

Ellen R, Harris H (2000) Introduction. Ellen RF, Parkes P, Bicker A (eds) Indigenous environmental knowledge and its transformations. Harwood, Amsterdam

Eyssartier C, Ladio AH, Lozada M (2008) Cultural transmission of traditional knowledge in two populations of North-western Patagonia. J Ethnobiol Ethnomed 4(1):25

Folkes C (2004) Traditional knowledge in social-ecological systems. Ecol Soc 9(3):7. http://www.ecologyandsociety.org/vol9/iss3/art7/. Accessed Febr 2010

Genin D, Simenel R (2011) Endogenous Berber forest management and the functional shaping of rural forests in Southern Morocco: implications for shared forest management options. Hum Ecol 39:257–269

Gerbati F (2004) La mobilisation territoriale des acteurs du développement local dans le Haut Atlas de Marrakech. Thèse pour l'obtention du Docteur de l'Université Joseph Fournier, Grenoble, France

Ghimire K, Pimbert M (1997) Social change and conservation. Earthscan Publications, London

Giannakopoulos C, Bindi M, Moriondo M, Tin T (2005) Climate change impacts in the Mediterranean. WWF, Gland, Switzerland

Grenier L (1998) Working with indigenous knowledge: a guide for researchers. International Development Research Centre, Ottawa

Guglielmino CR, Viganotti C, Hewlett B, Cavalli-Sforzas LL (1995) Cultural variation in Africa: role of mechanisms of transmission and adaptation. Proc Natl Acad Sci U S A 92:7585–7589

Haut Commissariat au Plan (HCP) (2007) Populations vulnérables: caractéristiques et implications. http://www.hcp.ma/. Accessed Oct 2009

Id Balkassm H (2002) Justice and indigenous Peoples in the world. Answers to the questions of the United Nations. Special Reporter in charge of following the violation of human rights and the rights of indigenous Peoples in the world. Justice and Amazigh People of Morocco. Tamaynut Association, Rabat, Morocco

Jellali M (1999) Large dams in Morocco: experience and perspectives for the future. Submission to the World Commission on Dams, Serial No: OPT063. Presented at the Africa/Middle-East Regional Consultation

Jodha NS (2007) Mountain commons: changing spaces and status at community levels in Hymalayas. J Mt Sci 4(2):124–135

Kadouiri A (2007) Les plantes aromatiques et médicinales dans les forêts de la vallée d'Agoundis (Commune d'Ijoukak, Province Al Haouz). Diversité et biomasse. Mémoire présenté à la Faculté des Sciences Semlalia pour l'obtention du Diplôme des Etudes supérieures approfondies. UFR: Biodiversité, biologie évolutive et paléontologie (BIOBEP)

Knippertz P, Christoph M, Speth P (2003) Long-term precipitation variability in Morocco and the link to the large-scale circulation in recent and future climates. Meteorol Atmos Phys 83: 67–88

Madhi M (1999) Pasteur de l'Atlas. Production pastorale, droit et rituel. Imprimerie Najah El Jadida, Casablanca

Ministry of Agriculture and Fisheries (2009) Green Moroccan Plan 2012–2020. http://www.ada.gov.ma/en/Plan_Maroc_Vert/plan-maroc-vert.php. Accessed May 2012

Minoia P, Brusarosco A (2006) Water infrastructures facing sustainable development challenges: integrated evaluation of impacts of dams on regional development in Morocco. The Fondazione Eni Enrico Mattei Note di Lavoro Series Index. http://www.feem.it/Feem/Pub/Publications/WPapers/default.htm. Accessed March 2012

Montanari B (2004) Alternative paths to commercialising medicinal and aromatic plants in a community of the High Atlas Mountains, Morocco. MSc Dissertation, University of Kent, Canterbury, UK

Montanari B (2012) A critical analysis of the introduction of essential oil distillation in the High Atlas of Morocco, with reference to the role of traditional gendered knowledge. Thesis submitted for the degree of PhD in Ethnobiology, University of Kent, UK

Neil Adger W, Huq S, Brown K, Declan C, Hulme M (2003) Adaptation to climate change in the developing world. Prog Dev Stud 3(3):179–195

Nusser M (2003) Political ecology of large dams: a critical review. Petermanns Geogr Mitt 147(1):20–27

Ostrom E (1990) Governing the commons: the evolution of institutions for collective action. Cambridge University Press, New York

Paine RT (1995) A conversation on refining the concept of keystone species. Conserv Biol 9:962–964

Parish R, Funnell DC (1999) Climate change in mountain regions: some possible consequences in the Moroccan High Atlas. Glob Environ Change 9:45–58

Pretty J, Pimbert M (1995) Beyond conservation ideology and the wilderness myth. Nat Resour Forum 19(1):5–14

Pretty J, Shah P (1997) Making soil and water conservation sustainable: from coercion and control to partnerships and participation. Land Degrad Dev 8:39–58

Russell DR (2004) Conservation, culture and development in the High Atlas Mountains of Morocco: making the case for collaborative management of natural resources at Toubkal National Park. Master of Arts Dissertation, Department of International Development Community and Environment, Faculty of Clark University, Worcester, Massachusetts

Saxena KG, Rao KS, Sen KK, Maikhuri RK, Semwal RL (2001) Integrated natural resource management: approaches and lessons from the Himalaya. Conserv Ecol 5(2):1–14

Schilling J Korbinian, Freier P, Hertige E, Scheffrana J (2012) Climate change, vulnerability and adaptation in North Africa with focus on Morocco. Agric Ecosyst Environ 156:12–26

Scoones I (1994) Living with uncertainty: new directions in pastoral development in Africa. I T Publications, London
Sillitoe P (2009) Local science vs global science: approaches to indigenous knowledge in international development. Berghahn books, New York
Sillitoe P, Bicker A (2004) Hunting for theory, gathering ideology. In: Bicker A, Sillitoe P, Pottier J (eds) Development and local knowledge. Routledge, London, pp 1–18
Sillitoe P, Marzano M (2009) Future of indigenous research in development. Futures 41:13–23
Smethurst D (2000) Mountain geography. Geogr Rev 90(1):35–56
Speth P, Fink AH (2010) Introduction. In: Speth P, Christoph M, Fink AH (eds) Impacts of global change on the hydrological cycle in West and Northwest Africa. Springer, Heidelberg, Dordrecht, London, New York
Tilt B, Braun Y, He D (2009) Social impacts of large dam projects: a comparison of international case studies and implications for best practice). J Environ Manag 90(3):S249–S257
Turner N, Garibaldi A (2004) Cultural keystone species: implications for ecological conservation and restoration. Ecol Soc 9(3):1
United Nations World Water Assessment Programme (2010) The implications of climate change on water. Highlights on climate change from the UN World Water Development Report 3: water in a changing world. http://unesdoc.unesco.org/images/0018/001863/186317e.pdf. Accessed May 2012

Climate Change and the Future of Mountain Farming on Mt. Kilimanjaro

Francis Mulangu and David Kraybill

1 Introduction

In the past 90 years, more than 80 % of the glaciers on Mt. Kilimanjaro in northern Tanzania have vanished due to climate warming. (Thompson et al. 2002; Kaser et al. 2004). The loss of these glaciers poses a critical threat to local farming, which for centuries has relied on glacial runoff as an additional source of irrigation for crops. Glacier melt, however, is not the region's only ominous climatic trend. In the past few years, major variations in precipitation have also been experienced in farming communities located in lower altitudes and on the outskirt of the mountain. Although total annual rainfall is not generally deficient, the yearly distribution of rainfall on the mountain is highly variable, thus posing a major threat to the heavily rain-dependent subsistent agricultural system. Average temperatures on the other hand have been increasing (see Fig. 1). Especially in lower altitudes, increase in average temperatures is further threatening the quality of Arabica coffee, Mt. Kilimanjaro's prime cash crop, that is already facing increasing rainfall variability. These climatic anomalies are so alarming that some scientists fear that the short rainy season (October–December) may soon disappear (Agrawala et al., 2003, Development and climate change in Tanzania: focus on Mount Kilimanjaro, unpublished).

In the midst of these growing climatic and food security threats, policymakers have been asking many questions regarding both the future of Mt. Kilimanjaro agriculture and the effective strategies for mitigating the threats of climate change. In fact, due to its volcanic soil, Mt. Kilimanjaro constitutes a major bread basket

F. Mulangu (✉)
African Center for Economic Transformation, Accra, Ghana
e-mail: mulangu@gmail.com

D. Kraybill
The Ohio State University, Ohio, USA

S. Mann (ed.), *The Future of Mountain Agriculture*, Springer Geography,
DOI: 10.1007/978-3-642-33584-6_6, © Springer-Verlag Berlin Heidelberg 2013

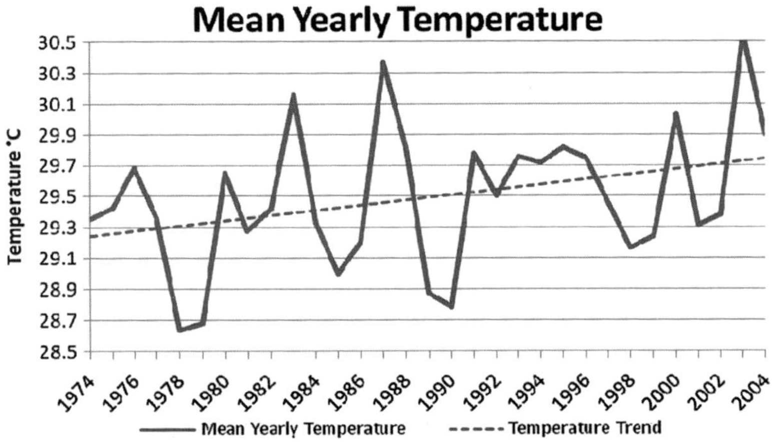

Fig. 1 Mean yearly temperature, Kilimanjaro region. *Source*

for more than one million people living in the Kilimanjaro region and regions surrounding it within Tanzania. A slight modification to its ability to produce agricultural products can have a number of serious impacts on food security. Therefore, rigorous economic analysis can contribute a great deal to the formulation of policy recommendations to help protect Mt. Kilimanjaro's agriculture from the potentially negative effects of climate change.

The present chapter uses a revenue function approach to estimate the impact of climate change on agriculture[1] on Mt. Kilimanjaro. The short rainy season is projected to suffer the most from climate change in East Africa (Agrawala et al., 2003, Development and climate change in Tanzania: focus on Mount Kilimanjaro, unpublished). For this reason, our study focuses on estimating the impacts of seasonal variations in precipitation on agricultural revenues. Since farmers in the region practice intercropping, the study developed a multivariate model that assumes interrelationships between crops. The rest of the chapter is organized as follows: We present an overview of agricultural activities on Mt. Kilimanjaro in Sect. 2. Section 3 presents the various methods used to estimate the impact of climate change. Section 4 presents the conceptual model that underlies our methodological approach. Section 5 discusses the data and fieldwork undertaken on Mt. Kilimanjaro. Sections 6 and 7 discuss the results and conclusion, respectively.

[1] Unlike weather, climate data are not always available, especially in developing countries. The difference between weather and climate is a measure of time. Weather is what conditions of the atmosphere are over a short period of time, and climate is how the atmosphere "behaves" over relatively long periods of time. The present chapter uses cross sectional data to establish the relationship between weather and crop revenue. This information is later used to make inferences about the impact of climate change on food security using weather to simulate climate change.

2 Agriculture on Mt. Kilimanjaro

Located 300 km south of the equator in the East African country of Tanzania, Mt. Kilimanjaro is the highest point in Africa with an altitude of 5,895 m. It consists of three peaks, Kibo, with an altitude of 5,895 m, Mawezi, with an altitude of 5,149 m, and Shira, with an altitude of 3,962 m. Mt. Kilimanjaro rises 4,800 m from the plains, covers four square km and at its widest is 40 km across (TANAPA 2004).

Seasonal rainfall distribution in particular dictates agricultural practices on Mt. Kilimanjaro. In the Kilimanjaro region, the year can be divided into four periods with respect to the amount of rainfall: There are two rainy seasons—a major one extending from March to May and a minor one from October to December. The major dry season lasts from January to February and a minor one from June to September. The amount of rainfall varies a great deal depending on altitude and the direction of the slope in the mountainous areas. The mean annual rainfall varies from 500 mm in the lowlands to over 2,000 mm in the mountainous areas (1,600 m and higher above sea level). Temperatures are closely related to altitude. During the rains, greater cloud cover and evaporative cooling tend to reduce maximum temperatures. Cloud cover also tends to raise minimum temperatures. The hot season lasts from October–March with high humidity; temperatures going up as far as 40 °C in the lowlands. In the mountainous areas temperature ranges from about 15–30 °C (URT 1998, 2002).

In general, the home parcels of Mt. Kilimanjaro farmers are characterized by an intensive integration of numerous trees and shrubs with food crops, cash crops, and livestock, simultaneously on the same unit of land. This setting enables farmers to obtain a sustain production with minimum external inputs. Soil fertility is enhanced by volcanic soil that contains a high base saturation and cation exchange capacity (Fernandes et al. 1985). Unfortunately, the steep slopes require intense erosion control procedures to reduce the soil loss due to torrential rains. Farmers practice multiple strategies to improve soil productivity, and improve erosion control. Our survey found that 46 % of farmers practiced bunds, 28 % practiced terracing, and 57 % practiced mulching in their home parcels.

Agroforestry is one common characteristic of Mt. Kilimanjaro farmers' home parcels. The parcels are constituted with multipurpose trees which are used to provide shade for coffee, to fence parcels, for fodder, mulching, and more. The main food crops produced are green plantain (hereinafter banana), beans, maize, cow pea, sweet potatoes, and tomatoes. The main cash crop is Arabica coffee, which is a perennial crop type and requires a lot of attention. The annual crops are generally produced in both rainy seasons, and the short rainy season harvest is a critical and strategic source of food between the long rainy seasons.

Irrigation is an important parts of Mt. Kilimanjaro agriculture. Although not all home parcels have access to irrigation, every home parcel has a network of irrigation furrows to tap and utilize run off from both the forest reserve of water above and rainfall. Our survey states that 28 % of home parcels are connected to an irrigation network. Most of the parcels are located in southern slope of the mountain. Due to the dry conditions, most villages located in the eastern slope of Mt. Kilimanjaro lack adequate irrigation networks. Consequently, farmers have developed methods such as shaping village paths to divert rain water toward their parcels.

3 Literature Review

Various approaches have been used to estimate the impact of climate change on agriculture. First, the agronomic approach uses agronomic models to simulate crop growth over the life cycle of the plant and measures the effect of climatic change on crop yield (Parry et al. 1999). One of the main advantages of the agronomic method is its inclusion of the effects of carbon fertilization in plants. This incorporation is valuable since failing to do so will overestimate the negative impact of climate change. On the other hand, critics of this approach argue that it tends to overestimate negative impact and underestimate positive impact because it fails to account for adaptations that farmers undertake in order to cope with climate pressures (Mendelsohn et al. 1994).

Other researchers have used Computational General Equilibrium (CGE) models to estimate the impact of climate change on economic outcomes (Roston 2003). CGE is an economy-wide model suitable for environmental issues, as it is capable of capturing complex economy-wide effects of exogenous changes while at the same time providing insights into the micro-level impacts on producers, consumers, and institutions. Usage of CGE has some drawbacks, nonetheless. Key limitations include the sensitivity of results to model assumptions, the weak empirical grounding of many of the model parameters, the absence of statistical tests for the model specification, and the complexity of the models (Gillig and McCarl, 2002, Introduction to computable general equilibrium models (CGE), unpublished).

The *Ricardian* approach is the mostly used approach to make inferences about the potential impact of climate change on agriculture. It was developed to account for farmers' adaptation (Mendelsohn et al. 1994). This approach uses cross-sectional data to capture the influence of climatic change on land value or farm income. The *Ricardian* approach is based on the notion that the value of a piece of land capitalizes the discounted value of all future profits that can be derived from the land. The model addresses climate change by including weather variables as determinants of land value. Last, since climate is a long run variable as opposed to weather, the *Ricardian* model in reality estimates the impact of weather variation on farm profit across space and makes inferences from the results about the potential impact of future climate change by simulating farm profit response to predicted future climatic change.

The present chapter uses a derivation of the *Ricardian* approach, the revenue function approach, to make inferences about the impact of climate change on food security. Although this approach shares similarities with the *Ricardian* approach, it differs from it as the revenue function approach used here breaks down farm revenue by crop type. This is important because unlike farmers in industrialized countries, smallholder farmers in developing countries simultaneously plant more than one crop type in the same parcel to reduce farm vulnerability to shocks. Failing to capture this portfolio diversification strategy would overestimate (underestimate) the negative (positive) impact of climate change.

Previous studies of climate change in Africa have analyzed spatial variation across vastly different localities to estimate climate impacts. A drawback of this approach is the potential bias from unobservable that varies across disparate regions and affects crop revenues. A fixed effect model can address this problem, but if the climate variables are fixed by locality, a fixed effect model will no longer be valuable since it will eliminate the weather variables from the regression. In contrast, this study uses the spatial variation in climate in a relatively small mountainous area with a high degree of ethnic homogeneity, similar crops, soil types, and access to markets, but marked differences in rainfall and temperature.

4 Estimating the Impact of Climate Change

4.1 Conceptual Model

We assume that each farmer seeks to maximize revenue subject to the exogenous conditions of the farm. Each farmer is assumed to choose a mix of agricultural techniques and inputs that achieve the highest crop revenue across all parcels, *j*, and crop types, *i*, such that:

$$rev_{ij} = \phi X_{ij} + \beta_1 R + \beta_2 R^2 + \theta H + \lambda I_i + \rho T + \gamma S_i + \varepsilon_{ij} \tag{1}$$

where rev_{ij} is per acre crop revenue, X_{ij} is a vector of inputs, R is a vector of a monthly rainfall, H is a vector of household characteristics, I_i is a dummy for irrigated land, T represents altitude which is used as a proxy for temperature, S_i is a vector of soil characteristics, and ε_{ij} is an error term which is assumed to be identically and independently distributed. Because this analysis is applied across multiple districts, it is quite possible that there are variables at the district level that are not taken into account in the analysis. Consequently, the study explores a district fixed-effects model that controls for district specific factors. Therefore, let's L represents a set of district dummies, Eq. (1) becomes:

$$rev_{ij} = \phi X_{ij} + \beta_1 R + \beta_2 R^2 + \theta H + \lambda I_i + \rho T + \gamma S_i + \eta L + \varepsilon_{ij} \tag{2}$$

The present chapter is interested in estimating the impact of climate change on farm welfare, here captured by farm revenue. In order to measure farm revenue response to precipitation change, the partial derivative of Eq. (2) is taken with respect to R_i, as shown in Eq. (3).

$$E\left[\frac{drev_{ij}}{dR}\right] = \beta_1 + 2\beta_2 E[R] \tag{3}$$

If Eq. (3) is positive (negative), the interpretation is that revenue increases (decreases) with precipitation increase.

In order to extrapolate the welfare value (G) of a change from today's climate (C_1) to future climate (C_2) n years from today, revenue before the change is subtracted from the revenue after the change for each farm household. The predictions of C_2 is taken from Hulme et al. (2001), who predicts 2020, 2050, and 2080 rainfall change in East Africa using a median of seven climate prediction models. The welfare change is the difference between revenues under the two climate scenarios. If the value is negative (positive), revenue declines (increases), leading to the conclusion that climate change would cause damage (benefit):

$$G = rev(C_2) - rev(C_1) \tag{4}$$

4.2 Empirical Model

Properly accounting for intercropping is not simple. Since more than one crop is cultivated in a plot, accounting for simultaneity among crop revenues implies a specification such as the system of equations below.

$$\begin{aligned}
rev_{coffee} &= \phi_1 X_{1j} + \beta_{1,1} R + \beta_{2,1} R^2 + \omega_1 W_j + rev_{maize} + rev_{banana} + rev_{other} + \varepsilon_{1j} \\
rev_{maize} &= \phi X_{2j} + \beta_{1,2} R + \beta_{2,2} R^2 + \omega_2 W_j + rev_{coffee} + rev_{banana} + rev_{other} + \varepsilon_{2j} \\
rev_{banana} &= \phi X_{3j} + \beta_{1,3} R + \beta_{2,3} R^2 + \omega_3 W_j + rev_{coffee} + rev_{maize} + rev_{other} + \varepsilon_{3j} \\
rev_{other} &= \phi X_{4j} + \beta_{1,4} R + \beta_{2,4} R^2 + \omega_4 W_j + rev_{coffee} + rev_{banana} + rev_{maize} + \varepsilon_{4j}
\end{aligned} \tag{5}$$

where rev_{coffee}, rev_{maize}, rev_{banana}, and rev_{other} represent crop revenues of coffee, maize, banana, and other, respectively, and $\omega W = \theta_1 H + \lambda_a I_i + \rho_a T_i + \gamma_a S_i + \eta_a L$ and the subscript a stands for equation number. Furthermore, irrigation usage could potentially be endogenous when estimating the revenue function. This is possible because of the value of crop revenue is enhanced by its proximity to irrigation canals. Failing to correct for endogeneity of irrigation may result in biased results. Therefore, we instrument irrigation usage with a dummy variable that is equal to one if a farmer's household supplies labor to maintain the communal irrigation system, and zero otherwise. This variable is a good instrument for two reasons. First, it correlated with irrigation usage because in every village, one must contribute to the maintenance of the irrigation scheme before lawfully using it. Second, revenue is not directly correlated with a farmer's contribution to the physical maintenance of the irrigation system.

Equation (5) is estimated using a Three Stage Least Squares (3SLS) model, so as to account for simultaneity arising from intercropping and to capture the error correlations between the revenue functions. Not all farmers sold the crops that they have produced. This is the case with major food crops such as maize. Failing to account for the sample selection bias may affect the results. We therefore use a corrective approach proposed by Shonkwiler and Yen (1999) to address it.

5 Data

The data used in the present chapter was collected under the auspices of the Kilimanjaro Livelihood and Climate Survey (KLCS) program. The KLCS started in September 2008 (Kraybill 2009). Funded by the Ohio State University Climate, Water, and Carbon (CWC) program, the project was a two-year mission to collect household level data on agricultural production, household socio-economic characteristics, and geographical coordinates of each household's location. Its mission was also to survey village level climatic variables, such as precipitation, on Mt. Kilimanjaro, Tanzania. The project surveyed fifteen villages, where fifteen households were randomly chosen in each village to make up a sample of 225 respondents. The 15 villages are located within the three districts that surround Mt. Kilimanjaro. The three districts are Hai, located on the western slope of the mountain, Moshi, located on the southern slope of the mountain, and Rombo, located on the eastern slope of the mountain (see Fig. 2). The northern slope of the mountain, partly located in Kenya, is mostly uninhabited and therefore does not figure in this study. The Kilimanjaro region has two other districts, Same and Mwanga. The present study did not include them because of their relative distant proximity to the mountain area.

Table 1 presents the socio-economic profile of the three districts. The Hai district is the largest with an area of 2,112 km^2, followed by Moshi and Rombo with an area of 1,713 and 1,442 km^2, respectively. Moshi has the largest population size of 505,287, followed by Rombo and Hai with population size of 417,602 and 200,136, respectively. The annual population growth rate is 1.3 % for Hai, 1.9 % for Moshi, and 2.4 % for Rombo. The average household size is the same for Hai and Moshi at 5.4 persons per household, but a little higher at 5.7 for Rombo. Finally, there are 33,899 households in Hai, 62,890 households in Moshi, and 35,078 households in Rombo.

The summary statistics of the data used in this chapter are presented in Table 2. On average, 76 % of farmers have applied manure in their parcels in both seasons. The average plot size is 1.4 acres with an average altitude of 1,303 m above sea levels (asl). In terms of input use, 28, 44, 50, 28, and 60 % of the respondent used pesticides, fertilizer, bunds, terracing, and mulching, respectively. 28 % of the respondent used irrigation, 10 % of the farmers paid for the irrigation and 20 % of them contributed to the maintenance of those irrigation facilities.

We aggregated the monthly precipitations into season such that: winter precipitations include the sum of precipitations in December, January, and February.

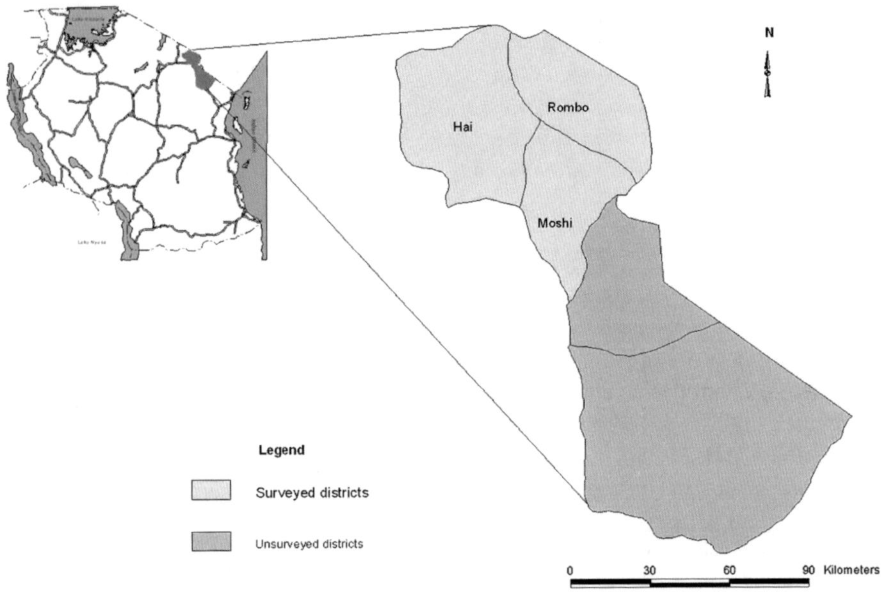

Fig. 2 Map of Kilimanjaro region showing the study area. *Source* www.wikipedia.org/wiki/Kilimanjaro_Region

Table 1 Kilimanjaro region (surveyed districts) demographics

District	Area	Administrative units			Population size	Annual population growth rate	Average HH size	No. of HH
		Division	Ward	Village				
Hai	2,112	4	14	82	200,136	1.3	5.4	33,899
Moshi	1,713	4	31	165	504,287	1.9	5.4	62,890
Rombo	1,442	5	20	62	417,602	2.4	5.7	35,078

Source URT (1998, 2002)

Spring precipitations include the sum of precipitations in March, April, and May. Summer precipitations include the sum of precipitations in June, July, and August. Fall precipitations include the sum of precipitations in September, October, and November. Using the rain gauges readings placed in all 15 villages, we found that average winter, spring, summer, and fall precipitation amounted to 130, 461, 104, and 238 mm, respectively.

The average percentage of land allocated to coffee, maize, and banana is 13, 32, and 34 %, respectively. The average revenue from coffee, maize, and banana is 14,181 TZ Shillings[2] (TZS) (TZS 7,520 in the short rainy season and TZS 21,433 in the long rainy season), TZS 4,751 (TZS 5,494 in the short rainy season and TZS

[2] At the time of the survey, \$1 = 1,300 TZ Shillings.

Table 2 Descriptive statistics

Variables	Mean	S. D.	Minimum	Maximum
Manure (total in kg)	0.76	0.42	0	1
Land size (acre)	1.4	0.85	0.17	4.1
Coffee revenue (TZ Shillings)	14,181	43,127	0	300,000
Maize revenue (TZ Shillings)	4,751	30,416	0	350,000
Banana revenue (TZ Shillings)	31,045	62,907	0	600,000
Altitude (m asl)	1,303	250	836	1,792
Winter precipitation (mm)	130	92	41	363
Spring precipitation (mm)	461	260	9	1,025
Summer precipitation (mm)	104	73	1.7	268
Fall precipitation (mm)	238	102	93	439
Pesticides (1 = used, 0 = not used)	0.28	0.45	0	1
Fertilizer (1 = used, 0 = not used)	0.44	0.5	0	1
Bunds (1 = used, 0 = not used)	0.5	0.5	0	1
Terracing (1 = used, 0 = not used)	0.28	0.45	0	1
Mulching (1 = used, 0 = not used)	0.6	0.5	0	1
Coffee share (in %)	13	16	0	60
Maize share (in %)	32	24	0	100
Banana share (in %)	34	25	0	100
Irrigation (1 = used, 0 = not used)	0.28	0.4	0	1
Irrigation Fee (1 = paid, 0 = not paid)	0.1	0.3	0	1
Irrigation maintenance (1 = paid, 0 = not paid)	0.2	0.4	0	1

Source Kilimanjaro livelihood and climate survey, 2008–2009

3,942 in the long rainy season), and TZS 31,045 (TZS 39,000 in the short rainy season and TZS 22,385 in the long rainy season), respectively. Note that in the analysis we included revenue from other crop, but we will not discuss it here because it lacks any policy relevance.

6 The Future of Agriculture on Mt. Kilimanjaro

6.1 Discussion of Empirical Results

Table 3 presents the main results of our analysis. For the coffee equation, we found that only fall precipitation has a statistically significant relationship with coffee revenue. This relationship is quadratic and U-shape. Winter, summer, and spring precipitations do not have a significant relationship with coffee revenue. Pesticides, altitude, and share of coffee have a positive relationship with coffee revenue.

In the maize equation, only fall precipitations do not have a significant relationship with maize revenue. Winter precipitations have a statistically significant relationship with maize revenue, and the relationship is quadratic and U-shape. Spring precipitations have a statistically significant relationship with maize

Table 3 Full model

	Coffee revenue		Maize revenue		Banana revenue		Other revenue	
Variables	Coefficients	S. E.	Coefficients	S. E.	Coefficients	S. E.	Coefficients	S. E.
Irrigation	−2.27	1.56	1.72*	0.82	−1.38	2.02	2.15	1.57
Pesticide	0.95**	0.51	0.37	0.21				
Manure	−0.45	0.66	0.52	0.36	2.53***	0.80	−0.25**	0.64
Bunds	0.35	0.48	−0.31	0.24	1.20***	0.45	−0.25	0.41
Terracing	−0.84	0.56	0.43	0.28	−1.62***	0.50	0.22	0.47
Mulching	−0.27	0.42	0.36	0.22	−0.46	0.49	0.06	0.39
Altitude	0.01***	0.00	−0.01***	0.00	0.01	0.00	−0.01**	0.00
Winter	9.56	18.51	−8.66***	10.00	−26.52	22.09	−15.09	17.78
Spring	−16.84	16.30	18.28**	8.45	−18.07	21.11	27.69*	15.46
Summer	10.95	12.98	−13.42**	6.75	8.70	16.88	−19.95*	12.48
Fall	−88.92***	22.41	30.05	13.11	−97.60***	24.31	50.35	21.26
Winter Sq	−1.04	2.00	0.92***	1.08	2.68	2.38	1.68	1.91
Spring Sq	1.21	1.22	−1.36**	0.63	1.59	1.55	−2.09**	1.14
Summer Sq	−0.98	1.34	1.34**	0.70	−0.95	1.73	2.03*	1.28
Fall Sq	8.41***	2.15	−2.77	1.26	9.11***	2.36	−4.76	2.03
Coffee Share	0.07***	0.01						
Maize Share			0.01***	0.00				
Banana Share					0.05***	0.01		
Other Share							0.00	0.01
Coffee Rev			0.25*	0.07	−0.15	0.17	0.36*	0.12
Maize Rev	1.16***	0.29			0.99	0.57	−0.92	0.44
Banana Rev	−0.42***	0.14	0.20**	0.07			0.26	0.12
Other Rev	1.03***	0.34	−0.23	0.15	1.38***	0.35		
Constant	230.52***	70.24	−84.59	38.97	346.49***	61.52	−134.90	64.22

() *p* value * $p < 0.10$, ** $p < 0.05$, *** $p < 0.01$
Source Author's estimates

revenue, and this relationship is quadratic and inverse U-shape. Summer precipitations have a statistically significant relationship with maize revenue, and the relationship is quadratic and U-shape. In addition, irrigation and maize share have a positive relationship with maize revenue while altitude has a negative relationship with maize revenue.

Banana revenue has a statistically significant relationship with fall precipitation only, and the relationship is quadratic and U-shape. Although not statistically significant, winter and summer precipitations seem to have an inverse U-shape relationship with banana revenue, while spring precipitations have a U-shape relationship with banana revenue. In addition, manure, bunds, and banana share have a positive relationship with banana revenue while terracing does not seem to be favorable for banana production.

There have been a number of criticisms of the Ricardian approach and approaches that derived from it over the years since it was first developed. There was initially a concern about the endogeneity of irrigation (Cline 1996; Schenkler et al. 2005). However, this study and other analyses (Kurukulasuriya and

Mendelsohn 2006; Mendelsohn and Seo 2007) now address this question carefully. Separate analyses on irrigated farms and rainfed farms were conducted and the results are presented in Tables 4 and 5, respectively. We found that coffee revenues from irrigated lands are responsive to summer precipitations while coffee revenues from rainfed lands are responsive to fall precipitations. Maize revenues from irrigated lands are responsive to all seasonal precipitations while maize revenues from rainfed lands are responsive to winter, spring, and summer precipitations only. Last, banana revenues from irrigated lands are not responsive to seasonal precipitations. However, banana revenues from rainfed lands are responsive to fall precipitations only. This heterogeneity in the crop responsiveness to rainfall to the presence or absence of irrigation implies that irrigation use is vital in helping farmers cope with the risks of climate change.

6.2 Climate Change and the Future of Farming on Mt. Kilimanjaro

To make predictions about the future of agriculture, we join the empirical results to future climate simulations. Hulme et al. (2001) predict change in rainfall by 2020 and 2050 based on 1961–1990 patterns. They present median of seven Global Climate Models (GCMs) experiments for both a slow global warming scenario and rapid global warming scenario. These GCMs include the CCSR-NIES (Emori et al. 1999), CGCM1 (Boer et al. 2000), CSIRO-Mk2., (Hirst et al. 2000), ECHAM4 (Roeckner et al. 1996), GFDL-R15 (Haywood et al. 1997), HadCM2 (Mitchell and Johns 1997), and NCAR1 (Meehl and Washington 1995). We use their result as they relate to East Africa, and simulate revenue response to the predicted change in precipitation while controlling for change in temperature.

Table 6 presents simulation results of the impact of climate change on crop revenue. For a slow climate change scenario, we predict that coffee revenue will decrease by an average of 18 and 28.8 % in 2020 and 2050, respectively. Maize revenue will increase by an average of 35.3 and 56.4 % in 2020 and 2050, respectively. And banana revenue will decrease by an average of 6.8 and 10.8 % in 2020 and 2050, respectively. Using a rapid climate change scenario, we predict that coffee revenue will decrease by an average of 42.5 and 63.1 % in 2020 and 2050, respectively. Maize revenue will increase by an average of 83.3 and 123.3 % in 2020 and 2050, respectively. Banana revenue will decrease by an average of 15.5 and 29.4 % in 2020 and 2050, respectively. However, the results concerning coffee and banana are not statistically significant.

The change in land use (crop allocation), crop yields, and revenues will mostly be influence by the extent to which climate has changed between now and in the future. Based on the results, it is fair to anticipate a gradual decrease in land allocation away from both banana and coffee, especially in lower altitude. In general, increased rain vulnerability will persuade farmers to take precautionary

Table 4 Irrigated lands only

	Coffee revenue		Maize revenue		Banana revenue		Other revenue	
Variable	Coefficients	S. E.	Coefficients	S. E.	Coefficients	S. E.	Coefficients	S. E.
Fertilizers			0.16	1.38				
Pesticide	−1.12	1.22						
Manure	−3.38***	1.12	0.40	1.33	4.15***	1.33	0.04	1.98
Bunds	−1.24	0.81	1.23*	0.71	0.92	0.91	1.43	1.04
Terracing	0.42	0.95	0.48	0.88	−0.77	1.14	2.03*	1.20
Mulching	0.95	1.49	−0.63	0.89	−2.96**	1.51	−3.93***	1.20
Altitude	0.01	0.01	−0.02**	0.01	0.00	0.02	−0.04	0.01
Winter	226.61	152.89	−327.09***	115.33	85.44	257.79	−602.40***	214.24
Spring	133.22	83.91	−213.41***	73.61	−8.74	178.38	−366.09***	156.22
Summer	−139.10*	80.65	228.21***	68.33	−4.16	172.60	346.80**	150.74
Fall	220.76	220.66	−320.79*	182.97	275.93	287.19	−584.55**	284.81
Winter Sq	−23.86	16.14	34.44***	12.18	−9.65	27.05	63.22**	22.53
Spring Sq	−11.43	7.04	17.93***	6.00	−0.05	14.76	31.14**	12.79
Summer Sq	14.12*	8.13	−23.04***	6.85	0.34	17.33	−34.70**	15.19
Fall Sq	−21.72	21.42	31.35*	17.73	−27.18	27.90	56.64**	27.63
Coffee Share	0.00	0.05						
Maize Share			0.05*	0.03				
Banana Share					0.08***	0.02		
Other Share							0.07**	0.03
Coffee Rev			0.21	0.42	0.22	0.31	−0.10	0.37
Maize Rev	0.74***	0.22			0.35	0.38	−0.79*	0.41
Banana Rev	0.14	0.25	0.20	0.13			−0.11	0.18
Other Rev	0.14	0.35	0.04	0.19	−0.04	0.43		
Constant	−1149.38	924.76	1668.59**	737.83	−811.35	1398.60	3167.67***	1255.52

() p value * $p < 0.10$, ** $p < 0.05$, *** $p < 0.01$
Source Author's estimates

measures by allocating more land toward the production of maize. This strategy is justified because higher CO_2 level will enhance the photosynthesis process in maize, thus improving its outputs. For this reason, we believe that maize competitiveness will increase to the detriment of coffee and banana in the region.

Because the approach used here makes estimations across space at a moment in time, the level of prices as a function of aggregate output is constant. The model cannot capture how prices would change if quantities of output were to change. However, we believe that climate change impacts on Mt. Kilimanjaro will not have significant impacts on price levels because the Kilimanjaro region is not big enough, globally speaking, for change in outputs in the region to have a large impact on price levels.

Table 5 Rainfed lands

	Coffee revenue		Maize revenue		Banana revenue		Other revenue	
Variable	Coefficients	S. E.	Coefficients	S. E.	Coefficients	S. E.	Coefficients	S. E.
Fertilizers			0.40*	0.24				
Pesticide	1.43*	0.78						
Manure	1.12	1.01	−0.25	0.43	3.72***	0.81	−1.26***	0.55
Bunds	−0.07	0.63	−0.27	0.26	0.49	0.55	−0.01	0.40
Terracing	−0.03	0.74	0.15	0.32	−0.60	0.76	−0.29	0.51
Mulching	−0.28	0.50	0.46**	0.22	−0.19	0.69	0.19	0.46
Altitude	0.01***	0.00	−0.01***	0.00	0.00	0.00	0.00*	0.00
Winter	16.40	17.60	−19.18***	7.88	−12.49	27.82	−14.45	19.45
Spring	−22.83	23.36	20.70**	9.89	−22.07	29.54	26.21	18.54
Summer	14.37	18.20	−14.73**	7.79	10.42	23.17	−17.73	14.92
Fall	−80.92***	26.51	20.06	12.44	−65.03***	25.45	22.80	21.16
Winter Sq	−1.86	1.89	2.09***	0.84	1.12	3.01	1.66	2.09
Spring Sq	1.70	1.77	−1.59**	0.75	1.95	2.21	−2.06	1.37
Summer Sq	−1.26	1.88	1.49*	0.80	−1.03	2.37	1.78	1.53
Fall Sq	7.86***	2.48	−2.00*	1.17	6.30***	2.35	−2.29	1.96
Coffee Share	0.09***	0.02						
Maize Share			0.01***	0.00				
Banana Share					0.04***	0.01		
Other Share							0.00	0.01
Coffee Rev			0.09	0.07	0.00	0.15	0.15	0.10
Maize Rev	1.00***	0.38			0.79	0.78	−0.70	0.52
Banana Rev	−0.54***	0.18	0.25***	0.08			0.28**	0.12
Other Rev	1.68***	0.63	−0.21	0.21	1.73***	0.49		
Constant	197.00**	84.39	−32.28	40.18	227.38***	88.60	−58.99	74.95

() *p* value * $p < 0.10$, ** $p < 0.05$, *** $p < 0.01$
Source Author's estimates

Table 6 Impact of climate change

Years	% Change Coffee revenue			% Change maize revenue			% Change banana revenue		
	Winter	Summer	Average	Winter	Summer	Average	Winter	Summer	Average
Slow scenario									
2020	−14.0	−22.0	−18.0	27.0**	43.5**	35.25**	−14.0	0.5	−6.8
2050	−22.4	−35.2	−28.8	43.2**	69.6**	56.4**	−22.4	0.8	−10.8
Rapid Scenario									
2020	−32.2	−52.8	−42.5	62.1**	104.4**	83.4**	−32.2	1.2	−15.5
2050	−60.2	−66.0	−63.1	116.1**	130.5**	123.3**	−60.2	1.5	−29.4

() *p* value * $p < 0.10$, ** $p < 0.05$, *** $p < 0.01$
Source Author's estimates

7 Conclusion and Policy Recommendations

This study used the revenue function approach to make inferences about the impact of climate change on the future of Mt. Kilimanjaro agriculture. Unlike previous studies, this study disaggregates both precipitation and crop revenue to capture the relationship between climatic variables and crop revenue while controlling for adaptation strategies made by smallholders through portfolio diversification.

Agricultural production on Mt. Kilimanjaro provides a unique setting to analyze the impact of climate change on future food production for two main reasons. First, variations in precipitation and temperature across altitudes provides a setting for natural experiments to estimate farm revenue's responses to climate conditions projected to occur in East Africa. Second, the area's farmers have similar cultures, employ similar agricultural methods, and are subject to similar market influences so the potential for bias is limited, given the relatively small geographical area they inhabit on and near Mt. Kilimanjaro.

Results provide evidence that maize revenue will increase by 35 and 56 % by 2020 and 2050. This is partly the case because increase CO_2 levels will improve the photosynthesis process of certain crop such as maize which will lead to higher yields. In addition, we suspect that higher temperature in higher altitude will transform certain farm to become more conducive for maize production as oppose to coffee production. Coffee and banana production will decrease. The water scarcity will particularly make lower altitudes coffee production more susceptible to pests which will further negatively affect yields and quality.

Policy Recommendations:

- *Designing farm management strategy around maize especially in lower altitudes*. Unlike the other crops, maize revenue is forecasted to increase with the anticipated climatic changes. Climate change mitigation strategies should be designed around maize. This will require developing new farm management strategy around and training extension officers to help facilitate the dissemination of the new maize-centered strategies such as planting nitrogen fixing legume as cover-crop to help increase soil moisture.
- *Subsidizing small scale irrigation systems in a market smart and sustainable way*. We found that access to irrigation will affect farm revenue responsiveness to seasonal precipitation. Considering the high cloud water retention capability of the thick forest of Mt. Kilimanjaro, climate change adaptation strategies for Mt. Kilimanjaro agriculture should focus on availing small-scale irrigation systems that will effectively reroute forest reserve water toward water reservoirs which in turn can be distributed to farms. This strategy must be drawn using maize as the main crop given the anticipated positive impacts of climate change on maize revenue. However, the way that subsidies are administered matters a lot. First, they should be designed to build markets by providing input vouchers that are redeemed through private input dealers. Second, the vouchers should be targeted as far as possible to those who do not currently use small scale

irrigation due to lack of credit, knowledge, or ability to withstand risks. Third, the subsidy should be modest in relation to other critical public expenditures such as a R&D and there should be a clear exit strategy.

- *Empowering and efficiently organizing local water-use-organizations.* Local water-use-organizations are pivotal for organizing water distribution and dealing with any water related conflicts in a community. These institutions can play the role of central planner by establishing an optimal community-wide schedule of water use by households or hamlets. However, in many cases water-use-organizations are plagued with low levels of capacity. These include inability to adequately coordinate water use throughout the community, low contribution from users, and inability to equitably solve local conflicts. Investing in these institutions can help improve water efficiency and reduce conflicts. Some possible solutions include: First, increase access to infrangible tools for helping these organizations effectively monitor water distribution and use. Second, the community needs to be involved in the maintenance and building of the water infrastructure and have a sense of ownership. Third, water use leaders must learn conflict management skills and be given access to local law enforcer to facilitate the settlement in more contentious cases.
- *The ministry of agriculture or equivalent entity should entrust a dedicated team to lead the design and implementation of climate change mitigation strategy.* One approach with some success in emerging countries such as Malaysia is to set up a dedicated and highly skilled transformation team from the government as well as leading outside experts, within the minister's office (or the equivalent). This team should have sufficient stature and experience to effectively dialogue with the private sector, including multinational firms. Coordination of donor efforts is also an important role for such a team. A high level agricultural transformation unit is in the process of being established in Ethiopia.

References

Boer GJ, Flato GM, Reader MC, Ramsden D (2000) Transient climate change simulation with greenhouse gases and aerosol forcing: projected climate to the twenty-first century. Clim Dyn 165:427–450

Cline W (1996) The impact of global warming on agriculture: comment. Am Econ Rev 86: 1309–1311

Emori S, Nozawa T, Abe-Ouchi A, Numaguti A, Kimoto M, Nakajima T (1999) Coupled ocean-atmosphere model experiments of future climate change with an explicit representation of sulfate aerosol scattering. J Meteorol Soc Jpn 77:1299–1307

Fernandes EC, Oktingati A, Maghembe J (1985) The *Chagga* homegardens: a multistoried agroforestry cropping system in Mt. Kilimanjaro (Northern Tanzania). Agrofor Sys 2:73–86

Haywood JM, Stouffer RJ, Wetherald R, Manabe S, Ramaswamy V (1997) Transient response of a coupled model to estimated changes in greenhouse Gas and sulphate concentrations. Geophys Res Lett 24:1335–1338

Hirst AC, O'Farrell SP, Gordon HB (2000) Comparison of a coupled ocean-atmosphere model with and without oceanic Eddy-induced advection. 1. Ocean spin-up and control integrations. J Clim 13:139–163

Hulme M, Doherty R, Ngara T, New M, Lister D (2001) African climate change: 1900–2100. Clim Res 17:145–168
Kaser G, Douglas RH, Thomas ML, Raymond SB, Tharsis MH (2004) Modern glacier retreat on Kilimanjaro as evidence of climate change: observations and facts. Int J Climatol 24:329–339
Kraybill D (2009) Kilimanjaro livelihood and climate survey. Website at http://digitalunion.osu.edu/R2/summer09/yoon/
Kurukulasuriya P, Mendelsohn R (2006) Modeling endogenous irrigation: the impact of climate change on farmers in Africa. Center for Environmental Economics and Policy in Africa (CEEPA) Discussion Paper No. 8. Special Series on Climate Change and Agriculture in Africa
Meehl GA, Washington WM (1995) Cloud Albedo feedback and the super-greenhouse effect in a global complex GCM. Clim Dyn 11:399–411
Mendelsohn R, Seo SN (2007) An integrated farm model of crops and livestock: Modeling Latin American agricultural impacts and adaptations to climate change. World Bank Policy Research Series Working Paper 4161. World Bank, Washington DC., USA, 41 p
Mendelsohn R, Nordhaus W, Shaw D (1994) The impact of global warming on agriculture: a Ricardian analysis. Am Econ Rev 84:753–771
Mitchell JFB, Johns TC (1997) On the modification of global warming by sulphate aerosols. J Clim 10:245–267
Parry M, Rosenzweig C, Iglesias A, Fischer G, Livermore M (1999) Climate change and world food security: a new assessment. Glob Environ Chang 9(1):S51–S56
Roeckner E, Arpe K, Bengtsson L, Christoph M, Claussen M, Dümenil L, Esch M, Giorgetta M, Schlese U, Schulzweida U (1996) The atmospheric general circulation model ECHAM-4: model description and simulation of present day climate. Max Planck Institute for Meteorology, Report No. 218, Hamburg
Roston R (2003) Modeling the economic impact of climate change. Working paper series N. 9, Ecological and Environmental Economics Program University of Venice, Italy
Schenkler W, Hanemann M, Fisher A (2005) Will U.S. agriculture really benefit from global warming? Accounting for irrigation in the hedonic approach. Am Econ Rev 95:395–406
Shonkwiler JS, Yen ST (1999) Two-step estimation of a censored system of equations. Am J Agric Econ 81(1999):973
Tanzania National Parks (TANAPA) (2004) Kilimanjaro: land, people, and history. African Publishing Group International, Harare
Thompson LG, Mosley-Thompson E, Davis ME, Henderson KA, Brecher HH, Zagorodnov VS, Mashiotta TA, Lin P-N, Mikhalenko VN, Hardy DR, Beer J (2002) Kilimanjaro ice core records: evidence of holocene climate change in Tropical Africa. Science 298:589–593
United Republic of Tanzania (URT), Planning Commission (1998) Kilimanjaro region socio-economic profile. Government Printers, Dar es Salaam
United Republic of Tanzania (URT), Planning Commission (2002) Kilimanjaro region socio-economic profile. Government Printers, Dar es Salaam

Is Mountain Farming No Longer Viable?

The Complex Dynamics of Farming Abandonment in the Pyrenees

Feliu López-i-Gelats

1 Introduction

In the last decades, particularly since the second half of the twentieth century, European mountains have gone through major socioeconomic and environmental transformations, as a consequence of accelerated integration within the national and international societal systems. This has not only led to an inundation of local cultures with external elements, but it has forced mountain regions to respond to wider social and economic trends, namely: full integration of rural economies into a globalised market economy; expansion of communication technologies; various population movements, both counter-urbanisation and out-migration; gradual environmentalisation of rural policy structures; increasing reconsideration of agricultural policy support; continued afforestation of agricultural land; emergence of new demands and interests on the rural space such as recreation, scenic beauty, biodiversity conservation and cultural museums, throughout a trend of patrimonialisation and commodification of the countryside; and growth of the service-based economy and rural gentrification (Vaccaro and Beltran 2007; Mitchley et al. 2006; Caraveli 2000; MacDonald et al. 2000; Ilbery 1998). The face of the European mountains is changing. Mountain populations are immersed in a process of accommodation of such rapid alterations, to face the arising uncertainties, to alleviate the new vulnerabilities, to take advantage of the potentialities coming up.

Certainly, all mountain regions have their own particular conditions and idiosyncrasies. Marked differences exist as regards their economic structures, their natural and human resources available, their peripherality from economic centres,

F. López-i-Gelats (✉)
Center for Agro-food Economy and Development (CREDA-UPC-IRTA),
Parc Mediterrani de la Tecnologia ESAB Building, C/Esteve Terrades, 8,
08860 Castelldefels Barcelona Spain
e-mail: feliu.lopez@upc.edu

S. Mann (ed.), *The Future of Mountain Agriculture*, Springer Geography,
DOI: 10.1007/978-3-642-33584-6_7,

their lack of infrastructures and social services, their topography, their weather conditions, and their demographic and social configurations. All these features strongly determine the nature and extent of the transition being experienced in each case. They all are going through the current changing times in a distinct and unique way. However, a common trend is identified all around the European countryside, namely the abandonment of mountain farming. In fact, the abandonment of farming, especially in marginal and less-productive lands, has become the most important trend in land-use and land-cover change in industrialised societies (Ramankutty and Foley 1999; Ilbery 1998; Moyano and Paniagua 1998; Baldock et al. 1996).

The decline of agropastoral activities is pervasive in the Pyrenees. A decrease in the number of farms, a reduction of the land devoted to farming, and a disregard of traditional farming practices (Garcia-Ruiz et al. 1996; MacDonald et al. 2000; Poyatos et al. 2003; Bartolomé et al. 2005; Mottet et al. 2006) jointly with a major shift in public attitudes from considering farmers as 'efficient producers of foodstuffs' to 'guardians of nature' or even 'unproductive suppliers' (Mitchley et al. 2006; Vaccaro and Beltran 2007; López-i-Gelats et al. 2011) are wide-ranging phenomena in the last decades. The vulnerability of the family-run transhumant livestock farming systems that characterize the Pyrenees is increasing under the on-going fast-changing circumstances. Mountain areas are especially vulnerable to change, and particularly to those outside-driven, as a consequence of their fragile ecosystems, local idiosyncrasies, small population size, remoteness and difficult accessibility, harsh weather and orographic conditions, poor and thin soils, lack of infrastructures, shortage of labour, as well as limited economies within the context of a globalized market. The agropastoral activity seems to be not any more the main pillar of the mountain's economy.

This chapter considers the critical situation farming in the Pyrenees now encounters, and understands it as part of a set of complex dynamics of change occurring between the environmental, economic and cultural domains. In particular, such transformations have to do with increasing trends of: (i) agroecosystem degradation, illustrated by processes of simplification of forage and pasture practices; (ii) economic restructuring, reflected by the high opportunity costs of all resources devoted to farming; and (iii) social recomposition, expressed by the changing role of farming and mountain areas as a whole.

In order to shed some more light on the abandonment of mountain farming, as follows we will revisit three specific researches (López-i-Gelats et al. 2009, 2010, 2011), which here are framed altogether to holistically examine the major trends that interplay in the abandonment of farming in the Pyrenees and thus to better understand the complexity and multifaceted nature of this process. Particular focus will be made in a specific region of the Pyrenees, the county of El Pallars Sobirà (42° 31′ 18″N and 1° 11′ 18″E), in Catalonia (Spain), at the very border between France, Spain and Andorra. This is a region of 1,378 km^2, and possesses an obvious mountainous character, which embraces large diversity of landscapes, extending from the hay meadows of the lower valleys and foothills into alpine highlands at peaks of around 3,000 m. Finally, in the light of these three works the rural development strategy of enhancing the multifunctionality of mountain farming as an option to cut its abandonment is examined.

2 The Complex Nature of the Abandonment of Farming in the Pyrenees

Abandonment of farming in the Pyrenees belongs to a complex dynamics established among various changes occurring in the environmental, economic and cultural spheres. The current context of high risk of abandonment is to a large extent the conjoint effect of the processes of (see Fig. 1): (i) agroecosystem degradation; (ii) economic restructuring; and (iii) social recomposition. The degradation of the mountain agroecosystems, that is, the simplification of forage and pasture practices, is a result of the adoption of low-cost and simplified management regimes (Poyatos et al. 2003; Mottet et al. 2006; López-i-Gelats 2010), as a consequence of the high opportunity costs that farming labour have to face (Laguna Marín-Yaseli and Lasanta 2003; García-Martínez et al. 2009), which mostly entails a neglect of the labour-intensive traditional farming practices. The restructuring of the mountain economy is due to the transition to a service-based economy, principally centred on the tourism sector (Laguna Marin-Yaseli and Lasanta Martinez 2003; Strijker 2005; García-Martínez et al. 2009; López-i-Gelats et al. 2011), and the growing difficulties for mountain farmers to remain competitive regarding the increased exposure to competition as a consequence of the full integration of mountain economies into the global market, with significantly lower income than their lowland counterparts (Tulla et al. 2003). Finally, the increased social recomposition triggers a changing in the role that more and more segments of society ascribe to farming and mountain areas as a whole. This is particularly due to the growing influence of urban and non-farming interests on these regions and their lifestyles (Vaccaro and Beltran 2007; López-i-Gelats et al. 2009), which brings about an augmented questioning of the food-producing determination of mountain farms and a rising demand for new goods and services, such as recreational activities, nature preservation, cheap housing, local culture, clean environment or animal welfare.

All these arising circumstances and social-ecological transformations are a mixture of external drivers and household farms' and local populations' adaptations. All of them reinforce each other and are leading towards a situation that makes that most Pyrenean household farms see it increasingly difficult to earn their livelihoods through farming. In view of this, the abandonment of farming is not, or at least not only, about farmers deciding to close down their farms and suddenly ceasing to undertake all the farming practices they had been routinely carrying out for decades day after day. The abandonment of farming is not, or at least not always, about farmland afforestation either. The abandonment of farming is not, or

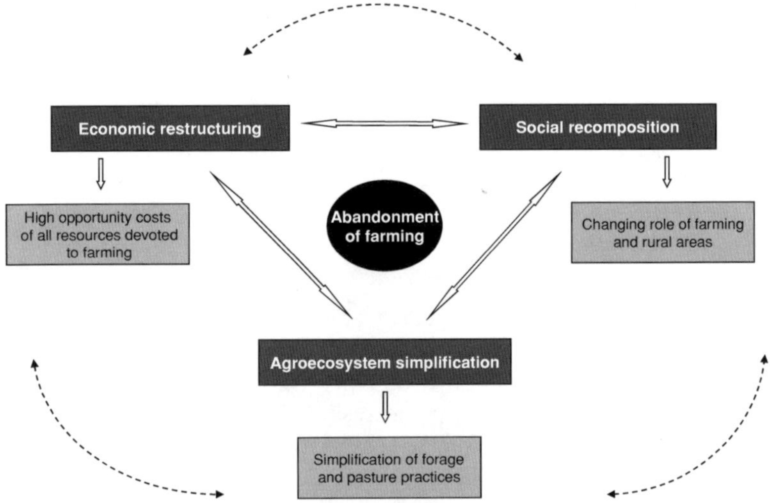

Fig. 1 The multifaceted nature of farming abandonment in the Pyrenees

at least not merely, about a shift from an utilitarian to a commodified conception of mountain regions. It should be seen as a sequence of household farm and discursive reorganizations to adapt to and frame the rapidly changing rural conditions, which turn out to be a more or less long process in which individual farms drop out little by little specific practices from their farming routines, and local people slowly shift the way to understand the role of mountain farming in society.

2.1 Agroecosystem Degradation: Simplification of Forage and Pasture Practices

Semi-natural grasslands are among the most species-rich habitats in the world, even comparable with rainforests (Wilson et al. 2012). As a consequence of the continual management through moderate mowing, grazing or fire, as well as the exclusive utilization of manure in fertilization, semi-natural grasslands maintain an enormous variety of grass species and wildflowers, as well as birds, insects and invertebrates. For instance, 65 % of the European Red List butterfly species depend on the maintenance of semi-natural grasslands (Van Swaay and Warren 1999). In Europe, 16 % of the total area of grassland is located in mountain regions (Sarzeaud et al. 2008). However, semi-natural grasslands are among the most threatened ecosystems. In fact, (FAO 2006) reports that the grassland areas in Europe decreased by 12.8 % from 1990 to 2003. The cessation of traditional labour-intensive livestock farming practices is the main factor threatening the preservation of mountain semi-natural grasslands.

Livestock farming in the Pyrenees is highly dependent on the availability of these semi-natural grasslands. It is characterised by the practice of transhumance, which entails a land-extensive management of the herd (sheep, cattle, horses and goats) between mountain hay meadows, which are located at medium-altitude lands, between 1,100 and 1,400 m a.s.l., for hay forage production for winter feeding (through both mowing and grazing), and the communal alpine pastures, between 2,200 and 3,000 m a.s.l., for spring and summer grazing, which constitute free feed sources.

Mountain hay meadows are a particular type of semi-natural grasslands, which in the case of the Pyrenees is dominated by the *Arrhenatherion elatioris* plant association. It is characterized by high abundance of hemicryptophyte graminoids, which are well adapted to diverse frequencies of mowing and grazing. These are characteristic habitats of Atlantic regions and only occasionally penetrate the most humid Mediterranean domains. In the Pyrenees, as pointed by Creus et al. (1984), the location of these hay meadows coincides with the isohyet of 900–1,000 mm of annual precipitation. As stated by Fillat et al. (2008), this indicates that these mountain hay meadows of the Pyrenees constitute the Southern distribution limit of this habitat in Europe. This implies that these Pyrenean semi-natural grasslands are particularly vulnerable to any type of change, such as water stresses and management modifications (Fillat et al. 1993). The conservation interest of this habitat is obvious. The conservation value of mountain semi-natural grassland is underlined by the Habitats Directive (Council Directive 1992), which considers mountain hay meadow a 'natural habitat type of Community interest' in representation of the Alpine biogeographic region.

However, in the last decades, farmers to accommodate the new circumstances arising and augment their viability are increasingly adopting low-cost, and less labour-intensive and simplified management regimes. These transformations are well reflected on the management of mountain hay meadows. The main strategy they follow is two-fold, on the one hand reducing costs, by means of abolishing expensive farming practices; and on the other, freeing labour force to be devoted to other economic activities by means of abolishing labour-intensive farming practices. In both cases, household resources previously devoted to farming are now redirected to other businesses, what entails a simplification of the livestock farming management regime. Consequently, this reduces the available time to undertake adequate farming practices, such as scrub clearance, adequate fertilisation, fencing and meadow approach road maintenance, appropriate management of the herd according to the forage resources available, etc. This gradual adoption of more and more low-cost and simplified management regimes is the process of increasing extensification of farming practices that takes place during the process of abandonment of mountain farming, which entails particular practices being regularly eliminated from the farming routine.

López-i-Gelats (2010) examined the effects of the gradual extensification on the species' composition and structure of *Arrhenatherion elatioris* mountain hay meadows. He considers two of the most representative adaptation strategies adopted by Pyrenean livestock farms: (i) the conversion of meadows into pastures, that is, from mown-and-grazed to only-grazed meadows (Garcia-Ruiz et al. 1996; Di Pietro 2001; Mottet et al. 2006); and (ii) the shift to forms of livestock farming more compatible with part-time farming, that is, from sheep- and cattle-farmed meadows to horse-farmed meadows (Laguna Marín-Yaseli and Lasanta 2003; Lasanta-Martínez et al. 2005; López-i-Gelats et al. 2009, 2011).

Thus, 164 samplings were conducted in fenced plots of farmed meadows to examine the vegetation' responses to partial abandonment. Each site was sampled by conducting 5-m linear transects laid out at random. All the species intercepted by a vertical pointer at 10-cm intervals were recorded along these lines. The plant cover was estimated using the Line-Intercept Method, adapted from Cummings and Smith (2000) by Sebastià (2004) for hay meadows in the Pyrenees. Plots were characterized by the determination of plant morphological and flowering traits, plant community traits, biodiversity indexes, production data, and fodder value. Finally, a Canonical Correspondence Analysis was conducted, by means of the CANOCO 4.5 program (Ter Braak and Šmilauer 2002).

López-i-Gelats (2010) observed that the only-grazed meadows and horse-farmed meadows are characterized by abundance of spring-flowering, very-short-flowering and rosulate hemicryptophyte species, that is, species commonly observed in semi-natural grassland communities under conditions of heavy grazing or under perturbation (Lavorel et al. 1997; Kahmen and Poschlod 2004; de Bello et al. 2005). Also, in these meadows higher proportions of forb, therophyte and uncommon species are identified. On the contrary, mown-and-grazed meadows, as well as cattle-farmed and sheep-farmed meadows, shows the features that are usually expected in well-preserved mountain hay meadows (Fanlo and Chocarro 1989; Chocarro and Reiné 2008; Lavorel et al. 1997), with abundance of legume, caespitose hemicryptophyte, scapose hemicryptophyte, Eurosiberian and very-long flowering species, as well as high values in hay forage production and quality.

The gradual simplification and extensification of farm managements, here illustrated by the dropping of mowing and the undertaking of forms of livestock farming more compatible with part-time farming, is manifestly modifying the species' composition and structure of mountain hay meadows. It goes with a degradation gradient of the habitat of mountain hay meadows, defined by a significant abundance of species not distinctive of mountain hay meadows and a remarkable loss in the main utility this habitat entails for the farming activity, that is hay forage production. The abundance of ruderal and species more characteristic of pasture habitats than hay meadows, as well as the low degrees of vegetation homogeneity and hay forage production, indicate that the mountain hay meadows under low-cost and simplified management regimes are immersed in a process of secondary succession towards other habitats than hay meadows with lower conservation and production interests. This result suggests that the preservation of semi-natural grassland requires more than simply guaranteeing the continuity of the livestock farming activity, but to maintain a set of adequate farming practices, such as mowing. Proactive policy measures to enhance full-time farming could play a prominent role.

2.2 *Economic Restructuring: High Opportunity Costs of All Resources Devoted to Farming*

The high age of farmers, the scarcity of labour, the low degree of mechanization, the lack of winter feeding, the increased competition from the outside market, the isolation from consumer centres, the harsh weather and orographic conditions, the local idiosyncrasies, the lack of infrastructures and public services, are among the reasons that put Pyrenean household farms in a situation of lack of viability and remarkable vulnerability to rapid changes (Baldock et al. 1996). In these conditions, to secure their livelihoods, household farms respond to the changing rural conditions through the uptake of specific farm adjustment strategies, particularly through either reorganizing the utilization of the farming resources available to adopt more viable agricultural management regimes or exploring new markets, which emerge from the new consumer and public policy demands on farming and rural areas, such as nature conservation, clean environment, animal welfare, recreation, etc. Mountain farms in the Pyrenees are immersed in a long-term process of reorganization.

Accordingly, agricultural land use in the Pyrenees is changing rapidly. Whereas the most productive farmland of the valley floors are being either used more and more intensively or displaced by infrastructures and tourism resorts (Laguna Marín-Yaseli and Lasanta Martínez 2003; Lasanta-Martínez et al. 2005), those less productive meadows located on steep slopes and at higher altitudes are being increasingly abandoned (Cernusca et al. 1996; Plieninger et al. 2006; Henle et al. 2008). This trend of polarization is being observed in mountain regions all around Europe (Gellrich et al. 2008; Gellrich and Zimmermann 2007; Cernusca et al. 1996). The way farming households are trying to secure their continuity is the chief driving force behind the ongoing transformations occurring in the Pyrenees. This has been also noted in the rest of Mediterranean mountain (Cialdella et al. 2009; García-Martínez et al. 2009; Fiorelli et al. 2007).

López-i-Gelats et al. (2011) explored the adjustment strategies undertaken by Pyrenean household farms. They conducted structured interviews to 22 % of the household farms of the county of El Pallars Sobirà and collected data on herd composition, land size and management, farm dynamism and continuity, labour and farmers' motivations for adopting nature protection measures. In doing so, a combination of Principal Components Analysis and Cluster Analysis was employed to explore farm characteristics and establish typologies of farms.

Four farm typologies are distinguished: from farms with remarkable availability of labour and land; to farms with wide availability of land, but where labour has been largely substituted by an increase in capital investment, reflected in higher rates of farm dynamism and intensification; also to farms with little access to land and labour, but that maintain a major capacity for capital investment, which is devoted to running farm tourism businesses and increasing farm mechanization; and, finally, to farms with limited availability of labour, land and capital. Similarly to what has been noted in other European mountains

(Cialdella et al. 2009; García-Martínez et al. 2009; Maye et al. 2009; Meert et al. 2005), the authors observed that each typology goes with the implementation of diverse farm diversification options: 'absence of diversification', when there is remarkable availability of labour and land; 'agricultural diversification', that is, the shift away from the production of traditional agricultural products towards the production of unconventional agricultural products, when there exists large availability of land; 'farmland diversification', that is, shifting away from the production of food, when there is little access to land and labour; and finally 'farm labour diversification', that is, the shifting of the family labour towards off-farm employment, when the household farm suffers from a very limited availability of labour, land and capital.

The different availability and allocation of farm resources—land, labour and capital—lie at the foundations of the different adjustment strategies followed by each type of farm. This has also been noted by other authors (Meert et al. 2005; Lobley and Potter 2004). It is argued that the four typologies identified reflect a gradation, as farm diversification practices are applied to more aspects of the farm household (labour, land, capital), the capacity of them to secure their livelihoods through farming decreases. That is, the spread of farm diversification practices is associated with a gradual marginalization of mountain farming: initially, abandoning traditional farming practices; secondly, abandoning the land resulting in high opportunity and management cost; finally, the very farming activity is abandoned. It is a process of transforming the nature of the farm household along which farms are gradually losing their food-producing determination. While the adoption of unconventional agricultural practices is associated with a better chance of agricultural continuity, this is not the case when the farm family members are mostly devoted to activities, such as farm tourism businesses or, especially, off-farm employment.

2.3 Social Recomposition: Changing Role of Agriculture and Rural Areas

In the last decades, a major shift in public attitudes, from considering farmers as 'efficient producers' to 'guardians of nature' or suppliers of multifunctional goods and services, is a wide-ranging phenomenon taking place all over the European mountains (Marsden 1995; McNally 2001; Mitchley et al. 2006). These regions are undergoing rapid processes of social recomposition and economic restructuring, which go with an increasing social complexity and new disputes about what is and should become the rural, as well as what should be the particular role of mountain farming. This situation is reflected in processes of value diversification and transformation, which currently are occurring in all Western societies (Marsden 1995; Cloke and Goodwin 1992; Wilson 2001; Jollivet 1997).

In line with this, (Murdoch and Pratt 1997) highlight the absence of a unique definition of the rural. Instead, the rural is defined through a range of phenomena that people experience to be and construct as rural. There is not thus an a priori definition of the rural, 'but rather a constellation of made, unmade, and remade constructions of the experience of it' (Lawrence 1997 p. 15). Accordingly, the current situation of mountain regions can be framed as battlefield where clusters of discourses of the rural struggle to impose their particular views and interests, their way of thinking and acting. Discourse analysis is thus an appropriate methodology to shed light on the current process of transformation of the role in society of both farming and rural areas. This is shown by the growing number of studies that in the last years have been devoted to this endeavour (Marsden 2008; Wolf and Klein 2007; Zografos 2007; Soliva 2007; Svendsen 2004; Elands and Wiersum 2001; Richardson 2000). Discourse analysis identifies the conditions behind a contested issue by means of uncovering the core assumptions, values and interests held by the various stakeholders. However, common perspectives on what constitutes the rural are often the result of a lack of attention paid to understanding the multiplicity of experiences and representations held by local residents about their own world (Halfacree 1995).

Thus, López-i-Gelats et al. (2009) attempt at taking a complete picture of the discourses on rurality held by the local population of the Pyrenees. Specifically, the multiplicity of representations and interests held by the inhabitants of the Pyrenean county of El Pallars Sobirà about their own world were identified by means of conducting a Q-methodology discourse analysis (Addams and Proops 2000; Zografos 2007; Hall 2008; Boonstra 2006; Brown 1980). They noted that the debate is organised by four discourses: conservationist, entrepreneurial, agriculturalist and endogenous development discourses.

According to the conservationist discourse, the main assets of mountain regions are its ecological attributes. 'We all need to believe that paradises do exist. This is one of the main roles of mountains' claimed a proponent of this discourse. For the advocates of this discourse, who are mainly urban incomers working in the public sector, it is essential for the future development of mountain regions to protect natural areas. For the entrepreneurial discourse, the main aim is to stop land abandonment, and for this the stimulation of economic growth becomes fundamental. The advocates of this discourse, who are mostly long-term residents working in the tourist industry, claim that the inherent lack of entrepreneurial initiative often prevents from taking appropriate advantages from the huge economic potential that mountain regions offer. Thus, one of the proponents of this discourse states: 'The best stake for the county is tourism. Tourism is an activity whose benefits are distributed evenly. The whole region benefits from it'.

The agriculturalist discourse argues that mountain livestock farming is an economic activity that goes beyond the mere production of meat and dairy products, and that it is well ingrained in the local culture. A proponent of this discourse states that 'a mountain without livestock is very sad'. The advocates of this discourse, who are mainly comprised by long-term residents devoted to the farming sector, highlight that many economic benefits, either directly or indirectly, derive from this activity. This discourse underlines the productive function of mountain farming and strongly opposes the role of farmers as 'guardians of nature'. They claim that are farmers the true 'endangered species'. Finally, the endogenous development discourse is concerned with the negative effects of the development strategy based solely on tourism. The advocates of this discourse, mostly returnees working in the public sector, claim for an economic development centred on listening to local population and enhancing diversification, by building bridges among the various economic sectors. A recurrent statement of the proponents of this discourse is: 'environmentalist, farming and tourist sectors all depend on the landscape, here. They should work together to prompt adequate development models' and 'the present tourism monoculture is dangerous. The economy of the county should be diversified'.

Specifically, on the future of mountain farming, the agriculturalist discourse argues that this activity guarantees many services that society welcomes, although it claims the production of foodstuff as the primary role of mountain farming. On the contrary, the entrepreneurial discourse sees mountain farming as a pre-modern occupation with no particular attributes, and highly uneconomical under the present conditions. The endogenous development discourse states that despite the central role farming deploys in the Pyrenees, this activity must be modernized and adapted to satisfy new societal demands, such as high-quality food, local culture and landscape preservation. Finally, the conservationist discourse fundamentally conceives farmers as guardians of nature.

López-i-Gelats et al. (2009) largely understand these discourses as an expression of an underlying social structure, based on the different experiences rural dwellers have of the changes undergoing, particularly: the rural population movements, both counter-urbanisation and out-migration; and the combined effect of the abandonment of farming and the tourism boom, that is, the tertiarisation of the local economy. The increasing presence of incomers and returnees to the detriment of long-term residents, jointly with the growing hegemonic role of tourism in the local economy at the expense of mountain farming, are the key processes that determine the fundamental differences in attitudes and interests among rural stakeholders and how they articulate them in discourses.

Rural mobility and job relocations are thus conceived as changes in locations or occupations entailing also shifts in the meanings ascribed to mountain farming and the rural as a whole. In practice, this reflects not only cultural differences on perceptions and values, but inequalities in opportunities. This is clearly shown by the fact that long-term residents and the people engaged in traditional forms of production, mostly proponents of the entrepreneurial and agriculturalist discourses, are the two segments of the population that complain most about being dispossessed by the on-going transformations. In contrast, the conservationist and endogenous development discourses, whose proponents are usually incomers and people who are not engaged in traditional forms of production, do not regret much the current transformations undergoing. It seems obvious that not only rural imaginaries but also socioeconomic reorganisations are in dispute.

3 Options to Cut the Risk of Abandonment of Mountain Farming in the Pyrenees

If 'mountain farming is no longer viable', what role, if any, should the farming activity play in the future development of the Pyrenees? In the policy debates on rural development, in the present situation of price–cost squeeze on farms, new emerging societal roles of mountain farming and degradation of forage and pasture resources, the development of mountain regions is generally associated with a shift away from the preoccupation with *economies of scale*, due to remoteness and physical disadvantages. Instead, it is predominantly advocated the dissemination of the multifunctional approach of *economies of scope*, that is, the promotion of a diversified rural economy, through increasing the range of the products and services provided to generate new economic activities and income sources allowing people to remain in mountain territories.

At present, the policy concept of multifunctionality is widely spread throughout the agro-food policy domain in Europe to cope with the structural weaknesses of rural and particularly mountainous areas (see McCarthy 2005 and López-i-Gelats and Tàbara 2010 for a genealogy of this policy notion). Mountain regions to guarantee the viability of the countryside must diversify their activities to aim at providing quality products, such as food safety, local products, rural and farm tourism, development of the heritage and landscapes and use of renewable energy. This concept is pervasive among rural policy-makers to suggest a strategy to cope with the current vulnerabilities of mountain farms. Multifunctionality has gained increasing relevance in debates on rural and agricultural development in the last years, particularly in marginal regions such as mountains (Knickel and Renting 2000; van der Ploeg et al. 2000; Pinto-Correia and Breman 2009; van der Ploeg et al. 2009; López-i-Gelats and Tàbara 2010). In fact, 'Enhancing the quality of life in rural areas and promoting diversification of economic activities' is one of the three major objectives of the EU Rural Development policy that has been set for the period 2007–2013 (Council Regulation 2005). Accordingly, the attainment of sustainable living mountains would involve a move away from the food-producing determination of mountain farm households to adopt new multifunctional rural development practices, such as off-farm employment, farm tourism businesses, nature conservation schemes or farm retailing.

Mountain farming is certainly multifunctional in nature. It is widely accepted that it provides more goods and services than industrial farming, e.g. nature conservation, forest fire prevention, reproduction of diversified rural landscapes or pathway maintenance. However, several approaches to the policy concept of multifunctionality coexist in both scientific and policy domains (Marsden and Sonnino 2008; Renting et al. 2009). In view of this, it is useful to distinguish between two different conceptions of multifunctionality: (a) the *multifunctionality of farmer*, and (b) the *multifunctionality of farming*.

The notion of the multifunctionality of farmer puts emphasis on enhancing the farmer's polyvalence, as the more convenient strategy of rural development, as also stressed by the entrepreneurial and conservationist discourses (Fig. 2). In practice,

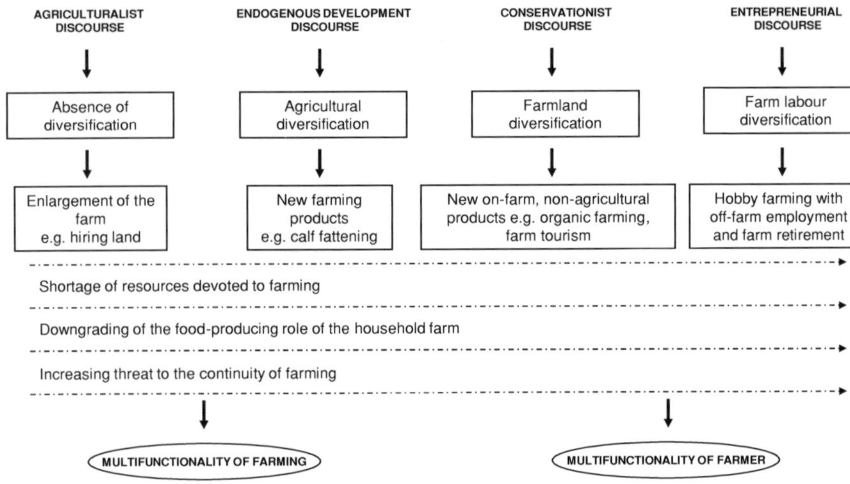

Fig. 2 Discourses, adaptation strategies and policy options dealing with the abandonment of mountain farming in the Pyrenees

this entails that farmers must behave as tourist operators, retailers, etc., while also farming. This endeavour can be successfully accomplished by farmers with a wide range of skills and resources. However, it is an unrealistic picture of mountain farming and development to ask farmers for performing systematically as *super-farmers*. In fact, this situation often ends up with a gradual disregard of labour-intensive farming practices, that is, favouring partial abandonment. On the contrary, the notion of the multifunctionality of farming stresses the multiple alternative social benefits that mountain farming directly and indirectly provides. Other mountain stakeholders—full-time tourist operators, full-time retailers, etc.—and sectors of the mountain economy will take advantage from the positive externalities mountain farming, particularly full-time mountain farming, brings about. The multifunctionality of farming puts emphasis, as also stressed by the agriculturalist and endogenous development discourses, on enhancing the synergies among different activities as the basis to reinforce rural development (Fig. 2). Accordingly, the economy of mountain regions should be diversified, but not at the expense of farming.

In the present circumstances of great risk of farming abandonment in mountain regions, and specifically in the Pyrenees, strategies of rural development centred on enhancing the multifunctionality of farmer encourage household farms to further adopt low-cost and simplified management regimes and shifting to other economic activities. This undermines the capability of mountain farming to make available the 'multiple' functions. On the contrary, strategies based on the consideration of the multifunctionality of farming guarantee more numerous and viable farms, as well as the provision of the other 'multiple' functions, through the preservation of the farming practices and resources (e.g. land, labour and livestock), and the consolidation of its food-producing determination. As a matter of fact, in contrast with the effects of enhancing the multifunctionality of farming, rural development strategies based on encouraging the multifunctionality of farmer result in an increasing alienation of the mountain farms from their social-ecological conditions, which lie at the foundations of their capacity of producing food and the social benefits that might be associated with the farming activity.

4 Conclusions

Certainly, the economy of mountain regions should be diversified to reduce the augmented vulnerability these regions suffer as a consequence of the accelerated changes undergoing, and to be in better conditions to face and even take advantage of the future uncertainties that undoubtedly will arise. However, from the integrative social-ecological perspective used here, it does not seem very effective that this should be done at the expense of the farming activity, given that this results in further social-ecological degradation, moving of farm resources away from agriculture or simply in substituting agriculture by other activities. As reasoned by Potter (2004), multifunctionality is an attribute of rural spaces, which facilitates that not only farmers but other actors than farmers can make a livelihood or benefit from the coexistence with farming activities. In this chapter, it is argued that in farming abandonment risk regions, as it is the case of the Pyrenees, it is possible to guarantee at the same time both viable farms and a diversified rural economy. However, this should be accomplished through the endorsement of rural development strategies based on the enhancement of the multifunctionality of farming, rather than the multifunctionality of farmer, that is, by means of encouraging those multifunctional activities that boost agriculture. In line with this, it is worth keeping in mind that what is multifunctional is not the farmer but farming.

References

Addams H, Proops J (2000) Social discourse and environmental policy: an application of Q methodology. Edward Elgar, UK

Baldock D, Beaufoy G, Brouwer F, Godeschalk F (1996) Farming at the margins: abandonment or redeployment of agricultural land in Europe. Institute for European Environmental Policy/ Agricultural Economics Research Institute, London

Bartolomé J, Plaixats J, Fanlo R, Boada M (2005) Conservation of isolated Atlantic heathlands in the Mediterranean region: effects of land-use changes in the Montseny biosphere reserve (Spain). Biol Conserv 122:81–88

Boonstra WJ (2006) Policies in the polder: how institutions mediate between norms and practices of rural governance. Sociol Rural 46:299–317

Brown SR (1980) Political subjectivity: applications of Q methodology in political science. Yale University Press, Yale

Caraveli H (2000) A comparative analysis on intensification and extensification in mediterranean agriculture: dilemmas for LFAs policy. J Rural Stud 16:231–242

Cernusca A, Tappeiuer U, Bahn M, Bayfield N, Chemini C, Filat F, Graber W, Rosset M, Siegwolf R, Tenhunan J (1996) ECOMONT: ecological effects of land use changes on European terrestrial mountain ecosystems. Pirineos 147–148:145–171

Chocarro C, Reiné R (2008) El cultivo de los prados en el Pirineo. In: Fillat F, García-González R, Gómez D, Reiné R (eds) Pastos del Pirineo. CSIC, Madrid, pp 141–158
Cialdella N, Dobremez L, Madelrieux S (2009) Livestock farming systems in urban mountain regions. Differentiated paths to remain in time. Outlook Agric 38:127–135
Cloke P, Goodwin M (1992) Conceptualizing countryside change: form post-Fordism to rural structured coherence. Trans Inst Br Geogr 17:321–336
Council Directive 92/43/EEC of May 1992 on the conservation of natural habitats and of wild fauna and flora
Creus J, Fillat F, Gómez D (1984) El Fresno de hoja ancha como árbol semi-salvaje en el Pirineo de Huesca (Aragón). Acta biologica montana 4:445–454
Cummings J, Smith D (2000) The line-intercept method: a tool for introductory plant ecology laboratories. In: Karcher SJ (ed) Tested studies for laboratory teaching 22. Proceedings of the 22nd conference of the Association for Biology Laboratory Education (ABLE), Clemson, SC, USA. Conference of the Association for Biology Laboratory Education, Clemson, 6–11 June 2000, pp 234–246
de Bello F, Lepš J, Sebastià MT (2005) Predictive value of plant traits to grazing along a climatic gradient in the Mediterranean. J Appl Ecol 42:824–833
Di Pietro F (2001) Assessing ecologically sustainable agricultural land-use in the Central Pyrenees at the field and landscape level. Agric Ecosyst Environ 86:93–103
Elands BHM, Wiersum KF (2001) Forestry and rural development in Europe: an exploration of socio-political discourses. For Policy Econ 3:5–16
Fanlo R, Chocarro C (1989) Influencia del 'efecto corte' sobre la variación florística y la producción de los prados de dalla: Pirineo Aragonés. Options Méditerr 3:341–344
FAO (Food and Agricultural Organisation of the United Nations) (2006) FAO statistical yearbook 2005–2006. FAO, Italy
Fillat F, Fanlo R, Chocarro C, Goded L (1993) Los prados de siega del Pirineo Central Español: Su función en el ciclo ganadero tradicional y perspectivas. In: Aguilera Sánchez JF (ed) Nutrición de Rumiantes en Zonas Áridas y de Montaña y su Relación con la Conservación del Medio Natural. Congresos y Jornadas 29/93, Sevilla: Junta de Andalucía, pp 15–34
Fillat F, García-González R, Gómez D, Reiné R (eds) (2008) Pastos del Pirineo. CSIC, Madrid 319 p
Fiorelli C, Dedieu B, Pailleux JY (2007) Explaining diversity of livestock-farming management strategies of multiple-job holders: importance of level of production objectives and role of farming in the household. Animal 1:1209–1218
García-Martínez A, Olaizola A, Bernués A (2009) Trajectories of evolution and drivers of change in European mountain cattle farming systems. Animal 3:152–165
Garcia-Ruiz JM, Lasanta T, Ruiz-Flano P, Ortigosa L, White S, González C, Martí C (1996) Land-use changes and sustainable development in mountain areas: a case study in the Spanish Pyrenees. Landsc Ecol 11:267–277
Gellrich M, Zimmermann N (2007) Investigating the regional-scale pattern of agricultural land abandonment in the Swiss mountains: a spatial statistical modelling approach. Landsc Urban Plan 79:65–76
Gellrich M, Baur P, Robinson BH, Bebi P (2008) Combining classification tree analysis with interviews to study why sub-alpine grasslands sometimes revert to forests: a case study from the Swiss Alps. Agric Syst 96:124–138
Halfacree KH (1995) Talking about rurality: social representations of the rural as expressed by residents of six English Parishes. J Rural Stud 11:1–20
Hall C (2008) Identifying farmer attitudes towards genetically modified (GM) crops in Scotland: are they pro- or anti-GM? Geoforum 39:204–212
Henle K, Alard D, Clitherow J, Cobb P, Firbank L, Kull T, McCracken D, Moritz RFA, Niemelä J, Rebane M, Wascher D, Young J (2008) Identifying and managing the conflicts between agriculture and biodiversity conservation in Europe—a review. Agric Ecosyst Environ 124:60–71
Ilbery BW (ed) (1998) The Geography of rural change. Longman, Harlow

Jollivet M (ed) (1997) Vers un rural postindustriel: rural et environnement dans huit pays européens. L'Harmattan, Paris
Kahmen S, Poschlod P (2004) Plant functional trait responses to grassland succession over 25 years. J Veg Sci 15:21–32
Knickel K, Renting H (2000) Methodological and conceptual issues in the study of multifunctionality and rural development. Sociol Rural 40:512–528
Laguna Marín-Yaseli M, Lasanta Martínez T (2003) Competing for meadows. a case study on tourism and livestock farming in the Spanish Pyrenees. Mt Res Dev 23:169–176
Lasanta-Martínez T, Vicente-Serrano SM, Cuadrat-Prats JM (2005) Mountain Mediterranean landscape evolution caused by the abandonment of traditional primary activities: a study of the Spanish Central Pyrenees. Appl Geogr 25:47–65
Lavorel S, McIntyre S, Landsberg J, Forbes TDA (1997) Plant functional classifications: from general groups to specific groups based on response to disturbance. Trends Ecol Evol 12: 474–478
Lawrence M (1997) Heartlands or neglected geographies? Liminality, power, and the hyperreal rural. J Rural Stud 13:1–17
Lobley M, Potter C (2004) Agricultural change and restructuring: recent evidence from a survey of agricultural households in England. J Rural Stud 20:499–510
López-i-Gelats F (2010) Are Mountains leaving agriculture behind? The complex dynamics of agricultural abandonment in the Pyrenees. PhD thesis. Barcelona: Institute of Environmental Sciences and Technologies (ICTA), Autonomous University of Barcelona
López-i-Gelats F, Tàbara JD (2010) A cultural journey to the agro-food crisis: policy discourses in the EU. J Agric Environ Ethics 23:331–344
López-i-Gelats F, Tàbara JD, Bartolomé J (2009) The rural in dispute: discourses of rurality in the Pyrenees. Geoforum 40:602–612
López-i-Gelats F, Milán MJ, Bartolomé J (2011) Is farming enough in mountain areas? Farm diversification in the Pyrenees. Land Use Policy 28:783–791
MacDonald D, Grabtree JR, Wiesinger G, Dax T, Stamou N, Fleury P, Gutierrez Lazpita J, Gibon A (2000) Agricultural abandonment in mountain areas of Europe: environmental consequences and policy response. J Environ Manage 59:47–69
Marsden T (1995) Beyond agriculture? Regulating the new rural spaces. J Rural Stud 11:285–296
Marsden T (2008) Agri-food contestations in rural spaces: GM in its regulatory context. Geoforum 39:191–203
Marsden T, Sonnino R (2008) Rural development and the regional state: denying multifunctional agriculture in the UK. J Rural Stud 24:422–431
Maye D, Ilbery B, Watts D (2009) Farm diversification, tenancy and CAP reform: results from a survey of tenant farmers in England. J Rural Stud 25:333–342
McCarthy J (2005) Rural geography: multifunctional rural geographies—reactionary or radical? Prog Hum Geogr 29:773–782
McNally S (2001) Farm diversification in England and Wales—what can we learn from the farm business survey? J Rural Stud 17:247–257
Meert H, van Huylenbroeck G, Vernimmen T, Bourgeois M, van Hecke E (2005) Farm household survival strategies and diversification on marginal farms. J Rural Stud 21:81–97
Mitchley J, Price MF, Tzanopoulos J (2006) Integrated futures for Europe's mountain regions: reconciling biodiversity conservation and human livelihoods. J Mt Sci 3:276–286
Mottet A, Ladet S, Coqué N, Gibon A (2006) Agricultural land-use change and its drivers in mountain landscape: a case study in the Pyrenees. Agric Ecosyst Environ 114:296–310
Moyano E, Paniagua A (1998) Agricultura, espacios rurales y medio ambiente. Revista Internacional de Sociología 19:127–152
Murdoch J, Pratt AC (1997) From the power of topography to the topography of power: a discourse on strange ruralities. In: Cloke P, Little J (eds) Contested countryside cultures: otherness, marginalisation, and rurality. Routledge, London
Council Regulation (EC) No 1698/2005 of 20 September 2005 on support for rural development by the European Agricultural Fund for Rural Development (EAFRD)

Pinto-Correia T, Breman B (2009) New roles for farming in a differentiated countryside: the Portuguese example. Reg Environ Change 9:143–152

Plieninger T, Höchtl F, Spek T (2006) Traditional land-use and nature conservation in European rural landscapes. Environ Sci Policy 9:317–321

Potter C (2004) Multifunctionality as an agricultural and rural policy concept. In: Brouwer F (ed) Sustaining agriculture and the rural environment. Governance, policy and multifunctionality, advances in ecological economics. Edward Elgar, Cheltenham, pp 25–35

Poyatos R, Latron J, Llorens P (2003) Land use and land cover change after agricultural abandonment: the case of a Mediterranean mountain area (Catalan Pre-Pyrenees). Mt Res Dev 23:362–368

Ramankutty N, Foley JA (1999) Estimating historical changes in global land cover: croplands from 1700 to 1992. Glob Biogeochem Cycles 13:997–1027

Renting H, Rossing WAH, Groot JCJ, van der Ploeg JD, Laurent C, Perraud D, Stobbelaar DJ, van Ittersum MK (2009) Exploring multifunctional agriculture. A review of conceptual approaches and prospects for an integrative transitional framework. J Environ Manage 90:S112–S123

Richardson T (2000) Discourses of rurality in EU Spatial Policy: the European spatial development perspective. Sociol Rural 40:53–71

Sarzeaud P, Pflimlin A, Perroc C, Becherel F (2008) Diversity of beef farming systems and grassland use in Europe. Grassland Sci Eur 13:693–705

Sebastià MT (2004) Role of topography and soils in grassland structuring at the landscape and community scales. Basic Appl Ecol 5:331–346

Soliva R (2007) Landscape stories: using ideal type narratives as a heuristic device in rural studies. J Rural Stud 23:62–74

Strijker D (2005) Marginal lands in Europe—causes of decline. Basic Appl Ecol 6:99–106

Svendsen GLH (2004) The right to development: construction of a nonagriculturalist discourse of rurality in Denmark. J Rural Stud 20:79–94

Ter Braak CJF, Šmilauer P (2002) CANOCO reference manual and conodraw for windows user's guide: software for canonical community ordination (version 4.5). Microcomputer Power, Ithaca, 500 p

Tulla AF, Soriano JM, Pallarès M, Vera A (2003) La transformación del model agrari en àrees de muntanya. ESPAIS 49:82–97

Vaccaro I, Beltran O (2007) Consuming space, nature and culture: patrimonial discussions in the hyper-modern era. Tour Geogr 9:254–274

van der Ploeg JD, Renting H, Brunori G, Knickel K, Mannion J, Marsden T, de Roest K, Sevilla-Guzmán E, Ventura F (2000) Rural development: from practices and policies towards theory. Sociol Rural 40:391–408

van der Ploeg JD, Laurent C, Blondeau F, Bonnafous P (2009) Farm diversity, classification schemes and multifunctionality. J Environ Manage 90:S124–S131

Van Swaay CAM, Warren MS (1999) Red data book of European butterflies (Rhopalocera). Nature and Environment, No. 99. Strasbourg: Council of Europe Publishing, 259 p

Wilson GA (2001) From productivism to post-productivism and back again? Exploring the (un)changed natural and mental landscapes of European agriculture. Trans Inst Br Geogr 26:77–102

Wilson JB, Peet RK, Dengler J, Pärtel M (2012) Plant species richness: the world records. J Veg Sci 23:796–802

Wolf SA, Klein JA (2007) Enter the working forest: discourse analysis in the Northern Forest. Geoforum 38:985–998

Zografos C (2007) Rurality discourses and the role of the social enterprise in regenerating rural Scotland. J Rural Stud 23:38–51

Future of Mountain Agriculture in the Alps

Christian Flury, Robert Huber and Erich Tasser

1 Introduction

In the European Alps, a total of roughly 18 % of the area is farmland and a further 18 % is covered by natural and semi-natural grassland (EEA 2005). In total, about 31.4 % of the Alps, with a surface area of 190,600 km^2 (Streifeneder 2010), are still used for agricultural purposes. Therefore, agriculture has a great responsibility for these areas and, by cultivating arable land, fields and pastures, makes an important contribution to the maintenance of a diversified cultural landscape in the Alps. However, there are marked differences between the eight Alpine states, Monaco, France, Switzerland, Italy, Germany, Liechtenstein, Austria and Slovenia: While almost half of the German Alpine area is used for agricultural purposes, in Italy only 25 % and in Slovenia less than 20 % (Tappeiner et al. 2008) is cultivated. From a proportional point of view, the largest Alpine areas are located in Austria (29 %), Italy (27 %) and France (21 %) (Streifeneder 2010).

Mountain agriculture in the European Alps has undergone considerable changes during the last 50 years. Compared to other mountain areas, such as the Andes or the Himalaya, the European Alpine area is influenced by four important factors which must be considered when analysing the development of mountain agriculture and its social significance: (1) The European Alpine regions are part of highly-

C. Flury (✉)
Agroscope Reckenholz-Tänikon Research Station, CH-8356 Ettenhausen, Switzerland
e-mail: christian.flury@art.admin.ch

R. Huber
Swiss Federal Institute for Forest,
Snow and Landscape Research WSL,
CH-8903 Birmensdorf, Switzerland

E. Tasser
EURAC Research, I-39100 Bozen, Italy

S. Mann (ed.), *The Future of Mountain Agriculture*, Springer Geography,
DOI: 10.1007/978-3-642-33584-6_8,

developed countries with, from a global point of view, an above-average degree of prosperity (IMF 2012). The tertiarisation is well advanced in the Alpine countries; on a country level the agricultural sector only holds a share in overall employment of between 2.4 (Germany) and 9.1 % (Slovenia) (PSA 2007). (2) The majority of the population in the European Alpine countries lives in cities or urban areas. In addition, there is a tendency for the population to concentrate in regional centres. The urban population and the fact that rural populations exhibit an increasing social orientation towards urban life have a significant influence on future development of mountain areas. In this context Messerli et al. (2011), even contend that the urban population, with its lifestyle and leisure activities, has a stronger influence than the population living in the Alpine region. At the same time, the mountain regions and agriculture are increasingly dependent on financial transfers and thus on the economic performance of the business centres (Weiss Sanpietro et al. 2004). (3) Due to the on-going economic development, agriculture is undergoing a progressive structural change accompanied by a reduction of the production factors labour and land. According to Schermer and Kirchengast (2006, p. 43), market integration of mountain agriculture leads to a switch of agriculture from a way of life to a form of business. Although rapid progress has been made in the mechanisation and rationalisation of mountain agriculture since World War II, the sector is unable to withstand the pressure of competition from more favourable locations due to the cost disadvantages resulting from its climatic, topographic and basic structural conditions (Schermer and Kirchengast 2006). (4) In the Alpine countries, the contribution to food security generated by agricultural commodities is of limited importance and is becoming smaller and smaller (Lauber 2006). Thus, the production potential of mountain agriculture is currently not fully exploited. This is in contrast to a more global perspective, since food security is at risk for 40 % of the mountain population worldwide (SAS 2012). The relatively low importance of production in the Alpine region can be explained by the switch from production designed to ensure self-sufficiency to market-orientated production which took place in the middle of the last century (Tasser et al. 2011).

In contrast, the multifunctional services provided by mountain farming in Alpine areas are becoming more and more important (Streifeneder 2010). The concept of multifunctionality is one of the central elements for the support for mountain agriculture in the European Union (Shucksmith et al. 2005), as well as in Switzerland (FOAG 2004) and Liechtenstein. At the same time, the optimisation of multifunctional services provided by mountain farming is a major objective of the mountain agriculture protocol of the convention on the protection of the Alps (Alpine Convention), which strives to achieve comprehensive protection and sustainable development of the Alps (Alpine Convention 1991). The objective to maintain and support the cultivation of traditional farmland and a form of agriculture which is compatible with both the location and the environment is based on the conviction that "by virtue of its wealth of natural resources, water resources, agricultural potential, historical and cultural heritage, value for quality of life and for economic and leisure activities in Europe and the transport routes crossing it, the Alpine region will continue to be of vital importance, particularly for the local population but also for the population of other regions," (Alpine Convention 1991, p. 2). In particular, the latter also covers tourism since, especially in industrialised countries, the mountain regions are important tourist destinations. In fact, die Alps are the world's second most important tourist region (SAS 2012) with over 540 million overnight stays per year.

This contribution explores the question of the development of mountain agriculture in the European Alpine region, both up till now and in the future, and the associated impact on the multifunctional services it provides. In the context of the specific situation and importance of the Alpine area, precedence is given to the ecosystem services which are incorporated into the overall concept of multifunctionality and which are linked to the utilisation and maintenance of open farmland as well as to the interaction between agriculture and the environment (Huber et al. 2012a). This contribution is designed as a meta-study and summarises the state of knowledge regarding the development and future of agriculture in the European Alps.

The answers to the questions raised are based on (a) a description of structural changes in agriculture in the Alpine area in recent years (Chap. 2); (b) a review of publications in the last few years on the subject of developments in mountain agriculture and the multifunctional services associated with land-use (Chap. 3); (c) as well as a compilation of model-based studies assessing the impact of (global) scenarios on land-use in the Alps (Chap. 4). Context-specific fields of action for mountain agriculture are identified, giving due consideration to future economic, natural and political basic conditions and the anticipated development in the Alpine area (Chap. 5).

2 Development of Agriculture in the Alps Since 1980

From a historical point of view, mountain agriculture and land-use in the European Alps has been undergoing continuous change for some considerable time (Siegl and Schermer 2012). From 1850 onwards, momentous political, economic and social changes accelerated the structural development of mountain agriculture. Between the beginning of the twentieth century and the end of the 1970s, the share of the population employed in agriculture sank in most Alpine areas from 70 to under 10 %. Over the last 30 years, changes have been modest when compared to the tertiarisation of the economy during this time. However, developments over this period show clearly the difference in the development of various regions with their specific natural and socio-economic characteristics. It is precisely these developments that form an important basis for the evaluation of short- and medium-term prospects for mountain agriculture in the Alps.

In spite of a wide range of regional, agricultural and environmental policy measures, the transition in mountain regions and mountain farming in the Alps has continued since the 1980s. Low incomes, a lesser degree of innovation in

Table 1 Structural change in the Alpine part of the countries

Country	Number of farms 1980	Number of farms 2000	Change of farms 1980–2000 per year (%)
Austria	109,554	96,205	−0.7
Switzerland	37,256	24,546	−2.1
Germany	29,041	22,017	−1.4
France	52,647	28,128	−2.9
Italy	165,607	93,046	−3.2
Liechtenstein	358	191	−3.1
Slovenia	53,089	23,149	−4.3
Alps	447,552	287,282	−2.2

Source Streifeneder et al. (2007)

comparison with other economic sectors, limited flexibility, tough global competition and the unfavourable topographic conditions cause potential successors to give up agriculture and mountain farms are abandoned. In some European regions, the high rate of abandonment has practically led to a collapse of agricultural structures and land-use, while in other regions structures and utilisation are only changing slowly.

Within the Alpine region there are significant differences at both regional and community level with regard to the development of farm structures, but also more particularly with regard to land-use. Although there are numerous local and regional peculiarities which influence the small-scale development of abandoned utilised areas and reforesting in suitable locations, favourable locations are less likely to be reforested than low-yield sites which require a lot of maintenance.

Viewed as a whole, between 1980 and 2000 the number of farms in the Alpine area went down by over a third, to 287,000 farms with at least one hectare of utilised agricultural land (Streifeneder et al. 2007). In particular, a great number of farm abandonments were recorded in a large part of the Italian Alps as well as in the French and Slovenian Alpine regions. On the other hand, the lowest number of abandonments was observed in the German and in particular the Austrian Alps (Table 1).

The trend to a declining number of farms continued in the decade up to 2010. Thus, for example, the number of farms in the Swiss mountain area sank by 1.7 % (Federal Statistical Office 2011) per year between 2000 and 2010, while in Austria the number of farms in the mountain area went down by 1.2 % per year between 1999 and 2010 (Statistics Austria 2012). Based on the evaluation of Streifeneder (2010) for the period 2000–2007, the decline in the number of farms in Italy, France and Germany in the decade up until 2010 is probably higher than in Switzerland and Austria. Regions with stable agricultural structures confront regions in the western and southern Alpine area where the dynamic structural development identified in 1980 continues unabated.

The abandonment of farms and the decline in the labour-force employed in agriculture lead to changes in land-use (Table 2). According to Streifeneder

Table 2 Development of utilised agricultural land in the Alpine region

Country	Utilised agricultural land 1980	Utilised agricultural land 2000	Change in utilised agricultural land 1980–2000 (%)	Change in number of farms 1980–2000	Agro-structural type Streifeneder et al. (2007)
Austria	1835369	1734369	−5.5	−12.2	Well performing region
Switzerland	805360	791938	−1.7	−34.1	Well performing region
Germany	511996	505433	−1.3	−24.2	Well performing region
France	84,9389	858650	1.1	−46.6	Uncorrelated structural change
Italy	1502027	1254044	−16.5	−43.8	Dynamic structural change
Liechtenstein	3634	3,593	−1.1	−46.7	Average structural change
Slovenia	210751	137566	−34.7	−56.4	Dynamic structural change
Alps	5718526	5285601	−7.6	−35.8	

Source Streifeneder et al. (2007)

(2010), the utilised area in the Alps sank by 7.6 % between 1980 and 2000. Slovenia and Italy had the highest decline in utilised area. In contrast, the utilised area of agricultural land in Liechtenstein, Switzerland and Germany remains practically stable or, in the case of Austria, falls slightly. However, there has been a clear drop in the utilisation of alpine pastures and unfavourable agricultural areas, as shown by the ongoing increase in reforestation (MacDonald et al. 2000). An investigation in the Swiss Alpine region reveals that new areas of scrub and forest are situated mainly on summer grazing pastures and in the uplands whereby this applies primarily to slopes or steep locations which involve a lot of labour or to poorly or undeveloped sites (Gellrich and Zimmermann 2007). The discrepancy between the decline in the agriculturally utilised area and the number of farms is a direct result of the marked increase in the average land area utilised by the remaining farms. As the size of the farms increases, the area farmed by one worker also increases which leads to a generally more labour-extensive form of farming concentrating more on those areas which are easily accessible and can be cultivated using machines. In regions with unfavourable agricultural production conditions, as much as two thirds of the previously utilised area is no longer worked,

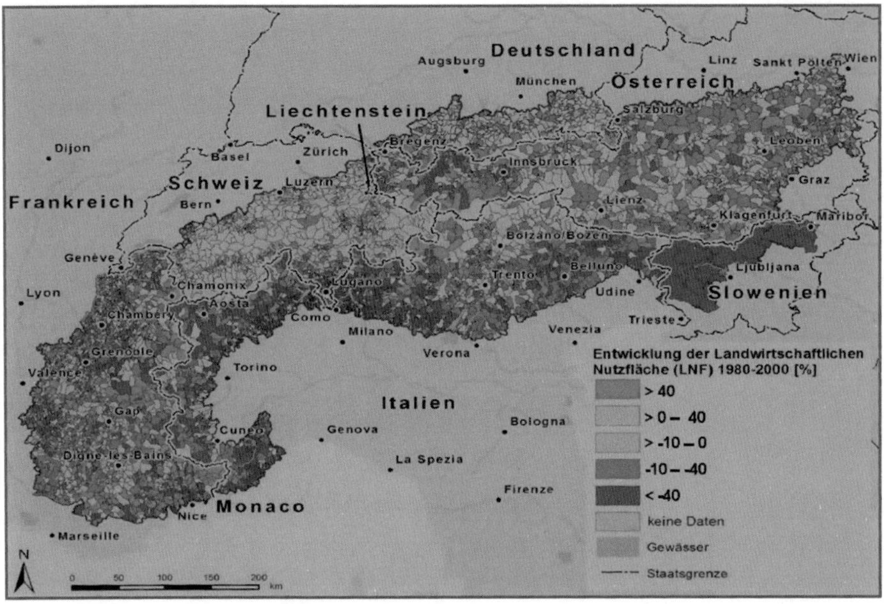

Illustration 1 Changes in agricultural utilised land in the Alpine area (1980–2000). *Source* Streifeneder 2010

while only very little land was taken out of cultivation in productive areas (Tasser et al. 2007). Insofar as land in marginal locations is not abandoned, it is only farmed extensively or used for grazing (Tasser and Tappeiner 2002; Pezzatti 2001) (Illustration 1).

The speed and pattern of changes in agricultural structures depends not only on the economic and social environment and existing farm structures, but also to a large extent on public support for mountain agriculture. While countries such as Italy (with the exception of the regions South Tyrol and Trento) and France have not attached any importance to the Alps for a long time, mountain areas and mountain farming have been receiving support in Germany, Austria and Switzerland (Bätzing 1996). Streifeneder et al. (2007) characterise these areas as "well performing regions" with moderate structural change and an assured income for mountain farms. Up till now, it has been possible to a large extent to prevent utilised agricultural land from being abandoned and left fallow in these countries. This is the result of favourable, economically viable structures and State support programs designed to offset location disadvantages plus farm income support. In spite of the fact that agricultural support in Liechtenstein is comparable to support provided in Switzerland, the structural change and decline in utilised area are both noticeably more pronounced. Streifeneder et al. (2007) typifies this development as "average structural change". By way of contrast, "dynamic structural change" can be observed in the development of mountain agriculture in Italy and Slovenia where a large number of farms have been given up and, compared to the Alpine region as a whole, an above-average amount of land has been abandoned. France occupies a special position since, according to the data on structural development harmonised by Streifeneder (2010), there has been no decline in the agricultural utilised land in the Alpine area in spite of the high number of farms abandoned ("uncorrelated structural change").

3 Literature Review on Agricultural Land-Use Change in the Alps

The concept of multifunctionality as a central element of mountain farming in the European Alps developed against the background of structural changes in agriculture and the ongoing marginalisation of the sector. Multifunctionality in general (Helming and Pérez-Soba 2011) and in mountain farming in particular is closely linked to agricultural land-use (Flury and Huber 2007). Cultivation and the intensity of the utilisation play a central role for the provision of services which go beyond the primary function of food production.[1] The maintenance of open landscapes (preservation of landscape vs. abandonment) is a vital factor in mountainous regions (MacDonald et al. 2000; Pointereau et al. 2008; Keenleyside and Tucker 2010). This is linked to various ecosystem services, which are of great importance to both the population in the Alps and society as a whole (Lauber 2006; Bacher et al. 2012; Huber et al. 2012a). In particular, this involves ecological services such as the maintenance of open cultural landscapes (Fischer et al. 2008; Lindemann-Matthies et al. 2011; Tasser et al. 2012b), the conservation of biodiversity (Tasser and Tappeiner 2002; Rudmann-Maurer et al. 2007), the protection of fertile land (Tappeiner and Cernusca 1998) or protection from natural hazards (Newesely et al. 2000; Tasser et al. 2003).

A comprehensive review of literature was carried out in order to be able to make a statement about the prospects for mountain agriculture and, by association, the provision of the multifunctional services. This involved the consultation of all those publications registered in Scopus between 2000 and 2012, with a title, abstract or key words containing the terms "land-use change" AND agriculture (farming) AND Alps (Alpine). Publications focusing strongly on non-agricultural aspects were rejected. This process resulted in 13 publications containing statements regarding the future of mountain agriculture. Table 3 contains a list of the publications consulted. In addition to the authors and the focus of the publication, the Table also includes the case-study region(s), the type of the agricultural change

[1] To a certain extent, the reduction of the idea of multifunctionality to land-use aspects leads to an incomplete view of the complexity of the actual concept. A comprehensive examination of multifunctionality in mountain agriculture would demand due consideration of a "territorial view" (Cairol et al. 2009), sustainability aspects (Renting et al. 2009) and the dynamic, temporal transition perspective (Wilson 2007). However, this would exceed the scope and objective of this Article.

(based on Streifeneder et al. 2007) and the methodological background. The central statements of the respective articles regarding future developments are cited in the last column.

The literature consulted covers the Alpine area quite widely and its diversity is well represented by various regions in France, Switzerland, Italy and also the Eastern Alps. In addition, the principal types of structural changes in agriculture as defined by Streifeneder et al. (2007) are covered completely. As various authors were studied, the analysis of the literature is a source of insight concerning the different points of view adopted by the respective institutions as well as the national attitude towards development, the status of mountain agriculture and the related country-specific problems. Agricultural land-use and its impact on multifunctional services and future structural change are investigated using statistic methods (e.g. Tasser et al. 2007) on the one hand, as well as with normative methods and also agent based land-use models (e.g. Briner et al. 2012) or linear programming models (e.g. Marini et al. 2011) on the other hand. One important characteristic of the more recent publications with an agro-economic orientation is that they combine agro-economic methods with other models and are thus able to assess changes in structures and land-use giving due consideration to climatic, ecological and socio-economic changes [e.g. (Gibon et al. 2010; Schreinemachers and Berger 2011; Briner 2012)].

The investigation of the literature resulted in the following conclusions concerning the future development of mountain agriculture, land-use and the associated maintenance of open cultural landscapes together with the related multifunctional services. They can be summarised under three headings:

- *Trends and existing driving force for land-use*. Some of the authors conclude that a continuing trend towards the abandonment of utilised land will have a negative impact on biodiversity and the associated services in the Alps and that existing measures are insufficient to counteract this change (Albert et al. 2008; Niedrist et al. 2009). In this context, Tasser et al. (2007) show that the former land-use intensity and proximity to forested areas play an essential role in the reforestation process of previously utilised agricultural land. Regardless of climatic change, the political and market environment will still continue to be the driving force behind the development of mountain agriculture in future (Briner et al. 2012).
- *Conservation of specific forms of utilisation*. Since specific types of vegetation can be attributed to certain forms of agricultural utilisation (Tasser and Tappeiner 2002), various authors conclude that specific forms of land-use should be conserved and encouraged in order to maintain the associated multifunctional services. Thus Quétier et al. (2007) call for conservation of mown meadows (mowing), Giupponi et al. (2006) consider the support of tradition livestock husbandry to be a key factor in the struggle to maintain open landscapes and Marini et al. (2009) identify extensive production systems and the utilisation of steep meadows as forms of agriculture which should be

Table 3 Literature on agricultural land-use in mountain regions and implications for the future

Authors	Focus	Region/Country	Structural change typology[a]	Methodology	Implication for the future of mountain agriculture
Albert et al. (2008)	Land-use change and tree dynamics	French Alps	Uncorrelated structural change	Land-use scenarios, based on habitat-suitability model and landscape model	Current management regime is not intense enough to resist colonization by larch in open and species rich grasslands. Ongoing and future agri-environmental policies have to be quickly adapted to protect biodiversity and ecosystem services provided by subalpine grasslands
Briner et al. (2012)	Modelling climate and land-use change	Visp/Switzerland	Well performing region	Modular modelling framework including forest, land-use and climate sub-models	Structural and economic trends will constrain agricultural production in Alpine regions despite potential favourable production conditions
Cocca et al. (2012)	Abandonment of livestock farming	Belluno Province/Italy	Dynamic structural change	Multiple regression models	Efforts are needed to maintain a territorial network of traditional extensive farms to avoid further landscape deterioration in Alpine areas
Gellrich and Zimmermann (2007), Gellrich et al. (2008)	Land abandonment and natural reforestation	Switzerland; four case studies in Switzerland	Well performing region	Multivariate statistical models; Classification analysis and interviews	General policy measures for the whole mountain area are not suitable for the prevention of land abandonment and forest re-growth and policy measures must pay more attention to local characteristics and needs
Giupponi et al. (2006)	Climate and land-use change, biodiversity and agri-environmental measures	Belluno Province/Italy	Dynamic structural change	Simple and multiple regressions analysis	Maintenance of the livestock production system typical of mountain agriculture is shown to be the key factor for contrasting land abandonment and the consequent expansion of woodlands

(continued)

Table 3 (continued)

Authors	Focus	Region/Country	Structural change typology[a]	Methodology	Implication for the future of mountain agriculture
Marini et al. (2009)	Impact of farm size (number of livestock units) on plant and insect diversity	Province of Trento/ Italy (tourism region)	Dynamic structural change	Multi-factorial mixed ANOVA	Regional stakeholders should consider targeted conservation schemes to prevent the ongoing substitution of small farms with large intensive farms (livestock units). (…) support farms with low production of organic fertilizers and reward the maintenance of the current management of steep meadows
Marini et al. (2011)	Mitigating the impacts of the transition from traditional to modern farming	Province of Trento/ Italy (tourism region)	Dynamic structural change	General Linear Mixed Models and Linear regression; Structured interviews with farmers	Agri-environmental schemes are positively related to both plant and insect diversity and might easily be implemented in future agri-environmental policy. However, we conclude that large and modern farms need to get more involved in biodiversity conservation as they will be the main actors in the future of Alpine farming
Monteiro et al. (2011)	Drivers behind loss of permanent meadows	Valtellina valley/ Lowlands Italy	Dynamic structural change	Land cover mapping; spatial bivariate analysis; GIS-based logistic regression model	In contrast to land use/land cover changes in other mountain regions; this study has found abandonment in the lowlands of Valtellina in addition to massive use intensification in the remaining meadows. Meadows habitat needs a well-designed landscape and farming planning (…)

(continued)

Table 3 (continued)

Authors	Focus	Region/Country	Structural change typology[a]	Methodology	Implication for the future of mountain agriculture
Niedrist et al. (2009)	Plant diversity	Bolzano/Bozen/Italy	Exceptional development	Discriminant analysis and biodiversity indices for 936 vegetation surveys	Number of plant communities along with the number of species decreases constantly and significantly with increasing land use intensity and on abandoned land. (…) Due to current trends, such as land abandonment and land use intensification, plant diversity in the Alps is decreasing considerably
Quétier et al. (2007)	Modelling ecosystem service sensitivity to land-use change	Villar d'Arène/Central French Alps	Uncorrelated structural change	Trait based modelling of plant functional types	Ecosystem services are most sensitive to changes in grassland management, supporting current agri-environmental policies aimed at maintaining mowing of subalpine grasslands in Europe
Tasser et al. (2007)	Land-use change and natural reforestation	Eastern Alps	Gradient of different dynamics	Historic photography, field work, Zero-inflated negative binomial model	Three most important driving forces of natural reforestation are the seed dispersal, the actual agricultural use and the years since abandonment. The less intensively the land was formerly used, the higher tree density in the reforested area
Zimmermann et al. (2010)	Land-use and land-cover change and biodiversity	European Alps (ecoregions)	Gradient of different dynamics	Hierarchical cluster analysis using historical maps and aerial photographs	The effect of land-use change depends on the landscape context. In other words, the same process (reforestation) has different effects in a largely agricultural area, compared with an area that already has a high proportion of forest

[a] Based on Streifeneder et al. 2007

encouraged. At the same time, larger farms should be definitely involved in the conservation of extensive forms of land-use (Marini et al. 2011).

- *More comprehensive solutions.* Basically, agricultural land-use in mountain regions is influenced primarily by a combination of local natural conditions and the basic conditions generated by economic and agricultural policy. However, comparable structural development processes or identical measures have different effects in different regions (Zimmermann et al. 2010). This leads to the realisation that if policy measures are not adapted specifically to the location, they are hardly likely to counteract the heterogeneity of abandonment of utilised land and forest encroachment in mountain areas brought about by natural and socio-economic conditions (Gellrich and Zimmermann 2007 and Gellrich et al. 2008). In addition, various authors place importance on the overall perspective. Monteiro et al. (2011) come to the conclusion that agriculture must be viewed within the context of a well-designed landscape planning. Cocca et al. (2012) reason that a territorial network is the only way of maintaining traditional land-use forms and the associated multifunctional services provided by agriculture.

On the basis of the literature studied, however, it is not possible to develop a direct causal relationship between regional differences in structural change (based on Streifeneder) and the policy implications drawn for the future of mountain agriculture. Indeed, the demand for the conservation of traditional agricultural systems is more pronounced in places where dynamic structural change is predominant. In the scientific debate, however, the same trends and policy measures are judged to be of importance, regardless of the nature of the structural changes in agriculture.

4 Scenarios for Agricultural Land-Use in the Alps

The major driving force behind changes in agricultural land-use in the Alps is of an economic and socio-cultural nature (Tappeiner et al. 2006; Briner et al. 2012). Therefore, the future of mountain agriculture depends strongly on overall economic and social developments. On an aggregated level, the future is characterised by numerous uncertainties. They can be structured consistently with the aid of scenarios to evaluate future developments. In this section, model-based studies assessing the impact of different scenarios on land-use in the Alps are summarised (Table 4).

Westhoek et al. (2006) developed four scenarios for rural regions in Europe. They apply to a time horizon up until 2030 and illustrate the relevant uncertainties for rural regions. At the same time, they are consistent with the climate scenarios of the IPPC (Abildtrup et al. 2006). The extent of globalisation (i.e. increasing integration of global markets vs. regional integration of markets) and regulation (little vs. high regulation) was chosen as the main driving force for future developments.

Table 4 Discussed scenarios of agricultural land-use

Source	Scale	Time horizon	Methodology	Scenarios
Van Meijl et al. (2006)	World (results focused on Europe)	2001–2030	General equilibrium modelling (GTAP) and land-use modelling (IMAGE)	Global economy (A1) Continental markets (A2) Global co-operation (B1) Regional communities (B2)
Tappeiner et al. (2006)	Stubaital (Innsbruck)	2033	Workshop; probabilistic simulation; optimization model	Status quo Environmentally friendly agriculture Support of regional economics
Nowicki et al. (2009)	Europe	2020	Macro-economic modelling on different scales (LEITAP, ESIM, CAPRI)	Reference scenario Conservative CAP scenario Liberalisation
Partidário et al. (2009)	Six case studies across Europe	Long term (not defined)	Scenario workshops with local stakeholders	Business as usual Managed change for biodiversity Liberalisation
Verburg and Overmars (2009)	Europe	2000–2030	Dyna-Clue, land use model	Global economy (A1)
Verburg et al. (2010)	Europe	2000–2030	Model linkage of GTAP, IMAGE, Dyna-Clue	Global economy (A1) Continental markets (A2) Global co-operation (B1) Regional communities (B2)
Mann et al. (2012)	Switzerland	2013–2020	SWISSLand (agent based sector supply model)	New Swiss agricultural policy 2014–2017
Renwick et al. (2013)	Europe	2020	CAPRI and Dyna-Clue	Removal of Pillar 1 and all market support WTO agreement and trade liberalisation Combination: removal of market support and trade liberalisation

- *A1 Scenario: Globalisation and little regulation*. The future is characterised by integrated global markets and relatively high economic growth. Agricultural productivity increases due to pronounced technological progress. Environmental problems are not regarded as a priority.
- *A2 Scenario: Continental markets*. In order to achieve the highest possible degree of independence in the field of food security, market solutions are sought primarily at regional level respectively with countries having the same values and standards. This is based on the maintenance of trade barriers and agricultural support. Cultural identity remains strongly anchored in the landscape and there is a minimum of State regulation.
- *B1 Scenario: Global co-operation*. Economic profits are coordinated at the international level by means of comparative cost advantages and trade in order to achieve wealth distribution, social justice and protection of the environment. This is based on far-reaching market liberalisation and State regulation to protect cultural values and the natural heritage.
- *B2 Scenario: Regional communities*. Local and regional communities become anchor points for society. The consumption of local products and self-sufficiency, responsibility for the environment and justice are the key to sustainable development. Far-reaching regulation and incentives for the conservation of small-scale farm structures are characteristics of a form of agricultural policy which is geared to self-sufficiency and ecological responsibility.

The scenarios have a range of implications for mountain agriculture. It is an accepted fact that the risk of the abandonment of utilised land is highest in mountain areas, and thus also in the Alps (Keenleyside and Tucker 2010). Utilisation of marginal land depends primarily on support for land-use generated by domestic support measures and to a lesser degree on the development of market access (Renwick et al. 2013). This is confirmed by earlier model results obtained by van Meijl et al. (2006), according to which the negative impact of liberalisation on land-use in Europe is, on the whole, low. Although results differ depending on the scenario, even in the worst case (Scenario B2) land-use does not decline much more than 10 %. In Scenario A2 (continental markets), there is even an increase in agriculturally utilised land. This is due to the fact that while the common agricultural policy (GAP) leads to an increase in extensive farming it does not result in abandonment of land. Model calculations carried out by Verburg and Overmars (2009) using the landscape model Dyna-CLUE concerning the explicit maintenance of open landscapes reveal that the abandonment of land is to be expected primarily in regions with unfavourable production conditions. Scenario A1 (global economy) shows that the decline in the Alps is comparable with that experienced by the Pyrenees, the Massif Central or the central German uplands (Mittelgebirge). While in three of the four scenarios the abandonment of utilised land continues to be the predominant change in land-use in Europe (Verburg et al. 2010), the degree to which it occurs varies from one Scenario to the other. In particular, the anticipated weaker level of economic growth in the EU15 is clearly apparent in the two B Scenarios where the abandonment of utilised agricultural land is much higher than in the two A

Scenarios with less State regulation. In fact, given high economic growth, the abandonment of utilised agricultural areas in Scenario A2 is supposed to be even lower than the expansion to new production locations. This is due to the fact that the European demand for food and energy must be satisfied from within Europe itself. Keenleyside and Tucker (2010) criticise the results of Verburg et al. (2010) and van Meijl et al. (2006) as unrealistic and, therefore, do not consider them in their conclusions. However, macro-economic effects cannot be disregarded when considering the future of mountain agriculture. Although the model Scenarios may seem unlikely, the modelled effect can nevertheless occur.

In addition to the Scenarios based on Westhoek et al. (2006), there are other model calculations dealing with the maintenance of open utilised agricultural areas in Europe (Keenleyside and Tucker 2010). Combined calculations using Dyna-CLUE and the agricultural sector model CAPRI reveal that the aggregated effects of liberalisation efforts would result in an overall abandonment of utilised agricultural areas of less than 10 % (Renwick et al. 2013), whereby there are considerable differences depending on the various types and sizes of the farms. In particular, there is a significant decline in grassland-based animal husbandry (especially sheep and goats) in the Mediterranean countries (−25 % UAA). On average, mixed livestock holdings likewise have no further need for over 10 % of their areas. Furthermore, it is primarily the small farms (<16 ha) which are abandoned. Combinations of small farms which produce milk or meat are frequently found in the Alps. Nowicki et al. (2009) show that a complete liberalisation would lead to a marked increase in the number of farms abandoned. A large share of these abandonments would be attributable to the effect of the discontinuation of direct payments.

The results of the various calculations reveal a high degree of heterogeneity between the individual regions. Different patterns of agricultural utilisation occur depending on macro-economic conditions and local natural and socio-economic characteristics (Verburg and Overmars 2009). This becomes very clear when smaller regions are used as units of investigation (Huber et al. 2012b). Partidário et al. (2009) base their evaluation of the future of rural areas on an inter- and transdisciplinary research design (Sheate et al. 2008) rather than on model calculations. Their qualitative evaluation of the future shows that in various case-study regions, and in particular in the Alpine regions, liberalisation scenarios have a negative impact on the fundamental indicators for sustainable development (Soliva 2007). However, a scenario which would promote the management and conservation of biodiversity is regarded as positive in all respects for sustainable land utilisation of rural regions. In addition, model calculations for Switzerland indicate that a switch from production-orientated payments for animal husbandry to area-based payments would have a favourable impact on the total utilised area in mountain areas (Mann et al. 2012). Tappeiner et al. (2006) use a variety of methods (trend analysis, stakeholder survey, agro-economic modelling) to evaluate the future of the Stubaital region near Innsbruck (Austria). Although local actors anticipate moderate changes in future, trend and modelling scenarios indicate a much more significant tendency towards the abandonment of utilised agricultural land. This divergence reflects the importance of the values and assumptions of those directly involved when evaluating future scenarios (Soliva and Hunziker 2009).

To summarise, the analysis of various propositions for agricultural land-use shows that the macro-economic and political environment is vital for the future development of mountain agriculture, that the impact is spatially heterogeneous and that there are both winners and losers. Therefore, the aggregation level plays an important role when evaluating the future of mountain agriculture in the Alps. On the whole, rural areas in general and mountain regions in particular are rated as losers in numerous scenarios, especially in the context of the reduction of direct payments. However, liberalisation of markets for agricultural products does not result in additional abandonments in every case. On the one hand, this reflects the shortage of the factor land which will gain in importance in the event of economic growth, increasing population and the associated rise in the demand for animal products. On the other hand, from a historic point of view, the degree of agricultural land utilisation has already sunk to a low level.

5 Discussion: Future of Mountain Agriculture in the Alps

The examination of future developments in mountain agriculture outlines the area of conflict in which this future will be situated. The fundamental demographic, economic and socio-cultural driving forces which were the basis for development in the past will continue to have a decisive influence on potential development paths over the next 20 years (Keenleyside and Tucker 2010). In this context, population dynamics and general economic developments will be major factors. From a historic point of view, these have been the cardinal causes of landscape change in the past (Siegl and Schermer 2012).

On the one hand, the model simulations discussed in Sect. 4 show that aggregated economic growth and the associated increase in demand is an important driving force behind land-use not only for European agriculture as a whole, but also for Alpine and mountain farming. With regard to the abandonment of utilisable agricultural land, the simulations imply that macro-economic developments could offset impacts of regional agro-environmental policies. This exemplifies that mountain agriculture is dependent on developments which cannot be influenced by itself. This situation will be intensified by the increasing influence of the urban population and the necessity for financial transfers to ensure the upkeep of fundamental services in peripheral regions (Messerli et al. 2011).

On the other hand, regional and local case-studies summarized in Sect. 3 indicate that economic pressure generated by increasing liberalisation accelerates structural development and specific forms of land-use are lost. Given natural and socio-economic characteristics, there is a danger that specifically areas of high nature value are lost (Keenleyside and Tucker 2010). This applies especially to locations where the rationalisation and mechanisation of mountain farming have reached their limits and the agricultural labour-force is no longer available (Tasser et al. 2012a). Consequently, the conclusion that the provision of multifunctional services by agriculture is no longer guaranteed leads to a demand for targeted support and the upkeep of the respective forms of land-use.

These two aspects reveal that there seems to be a contradiction between the aggregate (top-down) and the regional (bottom-up) point of view. While on the European level economic development stands in the foreground as the prime factor for full utilisation of agricultural land, the economic pressure at local level is rated as the driving force behind abandonment. The divergence in the views results from differing considerations regarding aspects of quality and quantity. In large-scale models the *quantitative* effects stand in the foreground while in individual case-studies the focus is often on *qualitative* und spatial-specific aspects.

Neither of these two aspects should be neglected when considering multifunctional agriculture. In the case of the abandonment of utilised land, the absolute area is not per se the decisive factor but rather the quality of the land abandoned, respectively of the land which is still cultivated. The provision of ecological services can be guaranteed using less land, but in spatially specific, or high-value locations. On the other hand, the promotion of traditional agriculture with small structures does not automatically guarantee the upkeep of the farms and their multifunctional services. Due also to the limited financial means available, macro-economic effects can more than offset the leverage of public support. The result is that while farmers may well remain within the sector, they earn the main part of their income in other branches and in spite of this they still abandon traditional utilisation forms. In addition, stable economic development is essential to ensure the availability of the financial resources without which it would be impossible to support peripheral regions in the first place.

Consequently, agriculture must adapt the provision of its multifunctional services to the societal demand (Lehmann 2002) and harmonise them in a spatially explicit manner (Grêt-Regamey et al. 2012). Regardless of the form of structural change which occurred in the past, maintenance and support of mountain agriculture can only be sustainable where a private or public service is provided for which there is a demand.

In addition, the fields of action for support of mountain agriculture depend on specific spatial and socio-economic characteristics. Insofar as overall economic development continues to permit utilisation of the land, targeted payments can assure the desired maintenance and support for the cultivation of traditional cultural landscapes and an environmentally-friendly form of agriculture in keeping with the location (Renwick et al. 2013). However, in the case of dynamic structural change it is possible that the same support payments will be ineffective (Huber et al. 2012b). It follows that measures to support sustainable mountain agriculture can only be effective and efficient if strategies are developed which are differentiated at a regional level (Rigling et al. 2012). Against this background, Schermer et al. (2012) developed specific options for action for various types of agriculture in the Tyrolean region.

To be able to make any use of endogenous growth potential, regional economic development is rated higher than sectoral agricultural measures in rural areas with declining employment and population (Dax 2001). On the other hand, selective agricultural support can produce the desired results in grassland areas with a high rate abandonment of summer grazing pastures. However, in grassland areas with a low rate of abandonment, measures must be envisaged to limit, in the first instance, any increase in intensity and to support diversification of production thereby ensuring that the potential of these regions can be exploited.

Better harmonisation between agriculture, nature conservation and development planning is viewed as a central field of action for farming in valley locations. In this case too, relationships within the agricultural sector are likewise a fundamental factor, in that diversification or growth in farm size can likewise lead to negative effects such as abandonment of utilisable land (Fischer et al. 2012). This means that there is also a need for action in the field of co-ordination within the agricultural sector and the evaluation of the distributional impact of policy measures supporting mountain farming. This represents a challenge for mountain agriculture which can only be overcome through dialogue with other users and providers of services (Lehmann and Messerli 2007). A cross-border exchange of experience could aid the efficient implementation of measures and tools (Bacher et al. 2012; Schermer et al. 2012).

In addition to the distributional impact, the heterogeneity of land-use, respectively multifunctional services also plays an important role (Rigling et al. 2012). In specific cases, forests likewise provide vital ecosystem services (carbon sequestration, protection from natural hazards), which under no circumstances should be neglected. In fact, Renwick et al. (2013) even identify a generally positive environmental effect resulting from the abandonment of utilised agricultural areas. On the other hand, climatic change becomes increasingly important when viewing the future. There is a great deal of uncertainty regarding the impact of climatic change (Beniston 2003) and it is possible that its effects will be small (Vittoz et al. 2009) or far less important than the socio-economic changes (Briner 2012). Nevertheless, mountain agriculture in the Alps will be faced with new challenges in the field of availability and the distribution of water resources (Beniston et al. 2011; Beniston 2012).

6 Conclusion

Basically, mountain agriculture will continue to be at a disadvantage due to its natural and structural production conditions. At the same time, mountain farming in the Alps has a heterogeneous appearance since it is characterised by numerous contrasts; it comprises valleys and summer grazing pastures, regional centres and peripheral valleys, production oriented areas and areas with low intensities, regions with high abandonment of marginal land and others with increasing intensive utilisation in favourable locations. This in turn also leads to heterogeneous prospects for the future. Thus, the fields of action for agriculture in the Alps must be differentiated at regional level to reconcile supply and demand of agricultural products and multifunctional services. This knowledge is not new and a large number of research results identify the need for differentiation in public support (Gotsch et al. 2004; Lehmann and Messerli 2007; Tasser et al. 2012b). However, it is of importance that macro-economic impacts should be taken into consideration when evaluating future regionally and locally beneficial measures. Otherwise there is the risk that fundamental economic developments will offset the impact of targeted measures designed to maintain and support the cultivation of traditionally utilised land and an environmentally-friendly form of agriculture which is in keeping with the respective location.

References

Abildtrup J, Audsley E, Fekete-Farkas M, Giupponi C, Gylling M, Rosato P, Rounsevell M (2006) Socio-economic scenario development for the assessment of climate change impacts on agricultural land use: a pairwise comparison approach. Environ Sci Policy 9:101–115

Albert CH, Thuiller W, Lavorel S, Davies ID, Garbolino E (2008) Land-use change and subalpine tree dynamics: colonization of Larix decidua in French subalpine grasslands. J Appl Ecol 45:659–669

Alpine Convention (1991) Convention on the protection of the Alps (Alpine Convention), Salzburg

Bacher M, Tasser E, Schermer M, Rudisser J, Tappeiner U (2012). Kulturlandschaft wohin? In: Tasser E, Schermer M, Siegl G, Tappeiner (eds) Wir Landschaftmacher. Vom Sein und Werden der Kulturlandschaft in Nord-, Ost- und Südtirol. Verlagsanstalt Athesia, Bozen, pp 187–212

Bätzing W (1996) Landwirtschaft im Alpenraum—unverzichtbar, aber zukunftslos? Eine alpenweite Bilanz der aktuellen Probleme und der möglichen Lösungen. Blackwell Wissenschafts-Verlag, Berlin

Beniston M (2003) Climatic change in mountain regions: a review of possible impacts. Clim Chang 59:5–31

Beniston M (2012) Impacts of climatic change on water and associated economic activities in the Swiss Alps. J Hydrol 412–413:291–296

Beniston M, Stoffel M, Hill M (2011) Impacts of climatic change on water and natural hazards in the Alps: can current water governance cope with future challenges? Examples from the European ACQWA project. Environ Sci Policy 14:734–743

Briner S (2012) Provision of ecosystem goods and services by agriculture and forestry in mountainous regions of Switzerland. Dissertation ETH, Zürich

Briner S, Huber R, Elkin C, Grêt-Regamey A (2012) Assessing the impacts of economic and climate changes on land-use in mountain regions: a spatial dynamic modeling approach. Agric Ecosyst Environ 149:50–63

Cairol D, Coudel E, Knickel K, Caron P, Kröger M (2009) Multifunctionality of agriculture and rural areas as reflected in policies: The importance and relevance of the territorial view. J Environ Policy and Planning 11(4):269–289

Cocca G, Sturaro E, Gallo L, Ramanzin M (2012) Is the abandonment of traditional livestock farming systems the main driver of mountain landscape change in Alpine areas? Land Use Policy 29:878–886

Dax T (2001) Endogenous development in Austria's mountain regions. Mt Res Dev 21:231–235

EEA (2005) Corine land cover 2000 (CLC2000) 100 m—version 5/2005

Fischer M, Rudmann-Maurer K, Weyand A, Stöcklin J (2008) Agricultural land use and biodiversity in the Alps. Mt Res Dev 28:148–155

Fischer M, von Felten S, Lauber S (2012) Heimfutterfläche—Schlüsselparameter der Sömmerungsnachfrage. Agrarforschung 3(4):194–201

Flury C, Huber R (2007) Evaluation der Jointness in der Schweizer Landwirtschaft. Agrarforschung 14:572–574

FOAG Swiss Federal Office for Agriculture (2004) Swiss agricultural policy: Objectives,tools, prospects. Bern

Gellrich M, Zimmermann NE (2007) Investigating the regional-scale pattern of agricultural land abandonment in the Swiss mountains: a spatial statistical modelling approach. Landsc Urban Plan 79:65–76

Gellrich M, Baur P, Robinson BH, Bebi P (2008) Combining classification tree analyses with interviews to study why sub-alpine grasslands sometimes revert to forest: a case study from the Swiss Alps. Agric Sys 96:124–138

Gibon A, Sheeren D, Monteil C, Ladet S, Balent G (2010) Modelling and simulating change in reforesting mountain landscapes using a social-ecological framework. Landsc Ecol 25:267–285

Giupponi C, Ramanzin M, Sturaro E, Fuser S (2006) Climate and land use changes, biodiversity and agri-environmental measures in the Belluno province, Italy. Environ Sci Policy 9:163–173

Gotsch N, Flury C, Kreuzer M, Rieder P, Heinimann HR, Mayer AC, Wettstein H-R (2004) Land- und Forstwirtschaft im Alpenraum—Zukunft im Wandel. Synthesebericht des Polyprojektes 'PRIMALP—Nachhaltige Primärproduktion am Beispiel des Alpenraums'. Wissenschaftsverlag Vauk, Kiel

Grêt-Regamey A, Brunner SH, Kienast F (2012) Mountain ecosystems services: who cares? Mt Res Dev 32:23–34

Helming K, Pérez-Soba M (2011) Landscape scenarios and multifunctionality: making land use impact assessment operational. Ecol Soc 16(1):50. http://www.ecologyandsociety.org/vol16/iss1/art50/

Huber R, Bebi P, Briner S, Bugmann H, Buttler A, Grêt-Regamey A, Hirschi C, Scholz RW, Zimmermann W, Rigling A (2012a) MOUNTLAND: Klimawandel und nachhaltige Landnutzung im Berggebiet. Agrarforschung 3(7–8):340–345

Huber R, Iten A, Briner S (2012b) Weiterentwicklung des Direktzahlungssystems: Auswirkungen auf die Landnutzung im Berggebiet. Agrarforschung 3(7–8):354–359

IMF International Monetary Fund (2012) World economic outlook database, April 2012. http://www.imf.org/external/pubs/ft/weo/2012/01/weodata/index.aspx

Keenleyside C, Tucker GM (2010) Farmland abandonment in the EU: an assessment of trends and prospects. Report prepared for WWF. Institute for European Environmental Policy, London

Lauber S (2006) Agrarstrukturwandel im Berggebiet. Dissertation ETH Nr.16716, Zürich

Lehmann B (2002) Multifunktionalität der Landwirtschaft aus ökonomischer Sicht. Agrarwirtschaft und Agrarsoziologie 57–68

Lehmann B, Messerli P (2007) The Swiss National Research Programme landscapes and habitats of the Alpine Arc. J Alp Res 4:19–28

Lindemann-Matthies P, Briegel R, Schüpbach B, Junge X (2011) Aesthetic preference for a Swiss alpine landscape: the impact of different agricultural land-use with different biodiversity. Landsc Urban Plan 98:99–109

MacDonald D, Crabtree JR, Wiesinger G, Dax T, Stamou N, Fleury P, Gutierrez Lazpita J, Gibon A (2000) Agricultural abandonment in mountain areas of Europe: environmental consequences and policy response. J Environ Manage 59:47–69

Mann S, Zimmermann A, Möhring A, Ferjani A, Mack G, Lanz S (2012) Welche Auswirkungen hat die Umlagerung der tierbezogenen Direktzahlungen? Agrarforschung 6(2012)

Marini L, Fontana P, Klimek S, Battisti A, Gaston KJ (2009) Impact of farm size and topography on plant and insect diversity of managed grasslands in the Alps. Biol Conserv 142:394–403

Marini L, Klimek S, Battisti A (2011) Mitigating the impacts of the decline of traditional farming on mountain landscapes and biodiversity: a case study in the European Alps. Environ Sci Policy 14:258–267

Messerli P, Scheurer T, Veit H (2011) Between longing and flight—migratory processes in mountain areas, particularly in the European Alps. Revue de géographie alpine/J Alp Res 99-1

Monteiro AT, Fava F, Hiltbrunner E, Della Marianna G, Bocchi S (2011) Assessment of land cover changes and spatial drivers behind loss of permanent meadows in the lowlands of Italian Alps. Landsc Urban Plan 100:287–294

Newesely C, Tasser E, Spadinger P, Cernusca A (2000) Effects of land-use changes on snow gliding processes in alpine ecosystems. Basic Appl Ecol 1:61–67

Niedrist G, Tasser E, Lüth C, Dalla Via J, Tappeiner U (2009) Plant diversity declines with recent land use changes in European Alps. Plant Ecol 202:195–210

Nowicki P, Goba V, Knierim A, van Meijl H, Banse M, Delbaere B, Helming J, Hunke P, Jansson K, Jansson T, Jones-Walters K, Mikos V, Sattler C, Schlaefke N, Terluin I, Verhoog D (2009) Scenar 2020-II—Update of analysis of prospects in the Scenar 2020 study—contract no. 30-CE-0200286/00-21. European Commission, Directorate-General Agriculture and Rural Development, Brussels

Partidário M, Sheate W, Bina O, Byron H, Augusto B (2009) Sustainability assessment for agriculture scenarios in Europe's mountain areas: lessons from six study areas. Environ Manage 43:144–165

Pezzatti M (2001) Einfluss der Erschliessung auf die Agrarstrukturen im Alpenraum. Eine agrarökonomische Analyse am Beispiel von vier Regionen in der Schweiz. Wissenschaftsverlag Vauk

Pointereau P, Coulon F, Girard P, Lambotte M, Stuczynski T, Sanchez Ortega V, Del Rio A (2008) Analysis of farmland abandonment and the extent and location of agricultural areas that are actually abandoned or are in risk to be abandoned. European Commission Joint Research Centre. Institute for Environment and Sustainability

PSA Permanent Secretariat of the Alpine Convention (2007) Transport and mobility in the Alps. Report on the State of the Alps, Innsbruck

Quétier F, Lavorel S, Thuiller W, Davies I (2007) Plant-trait-based modeling of ecosystem service sensitivity to land-use change. Ecol Appl 17:2377–2386

Renting H, Rossing WAH, Groot JCJ, Van der Ploeg JD, Laurent C, Perraud D, Stobbelaar DJ, Van Ittersum MK (2009) Exploring multifunctional agriculture. A review of conceptual approaches and prospects for an integrative transitional framework. J Environ Mt 90:112–123

Renwick A, Jansson T, Verburg PH, Revoredo-Giha C, Britz W, Gocht A, McCracken D (2013) Policy reform and agricultural land abandonment in the EU. Land Use Policy 30:446–457

Rigling A, Huber R, Bebi P, Brand F, Briner S, Buttler A, Elkin C, Gillet F, Grêt-Regamey A, Hirschi C, Lischke H, Scholz RW, Seidl R, Spiegelberger T, Walz A, Zimmermann W, Bugmann H (2012) Sustainable land use in mountain regions under global change: synthesis across scales and disciplines. Ecol Soc (in revision)

Rudmann-Maurer K, Weyand A, Fischer M, Stöcklin J (2007) Microsatellite diversity of the agriculturally important Alpine grass Poa alpina in relation to land use and natural environment. Ann Bot 100:1249–1258

SAS (2012) Swiss Academy of Science: Die Berggebiete—eine Priorität für eine Welt unter Druck und für die Schweiz. Faktenblatt Rio + 20 Nr. 2, Bern

Schermer M & Kirchengast C (2006) Perspektiven für die Berglandwirtschaft. In: Psenner R, Lackner R. (Hg.): Die Alpen im Jahr 2020. Innsbruck: innsbruck university press (IUP) 41–55

Schermer M, Siegl G, Steinbacher M, Tappeiner U, Tasser E (2012) Strategien für die Kulturlandschaft der Zukunft. In: Tasser E, Schermer M, Siegl G, Tappeiner U (eds) Wir Landschaftmacher. Vom Sein und Werden der Kulturlandschaft in Nord-, Ost- und Südtirol. Verlagsanstalt Athesia, Bozen, pp 223–245

Schreinemachers P, Berger T (2011) An agent-based simulation model of human-environment interactions in agricultural systems. Environ Model Softw 26:845–859

Sheate W, Partidário M, Byron H, Bina O, Dagg S (2008) Sustainability assessment of future scenarios: methodology and application to mountain areas of Europe. Environ Manage 41:282–299

Shucksmith M, Thomson KJ, Roberts D (2005) The CAP and the Regions: the TerritorialImpact of the Common Agricultural Policy. CAB International, London, Washington

Siegl G, Schermer M (2012) Kulturlandschaft woher? In: Tasser E, Schermer M, Siegl G, Tappeiner U (eds) Wir Landschaftmacher. Vom Sein und Werden der Kulturlandschaft in Nord-, Ost- und Südtirol. Verlagsanstalt Athesia, Bozen, pp 57–103

Soliva R (2007). Agricultural decline, landscape change, and outmigration. Mt Res Dev 27
Soliva R, Hunziker M (2009) Beyond the visual dimension: using ideal type narratives to analyse people's assessments of landscape scenarios. Land Use Policy 26:284–294
Statistics Austria (2012) Agrarstrukturerhebung 2010. Wien
Streifeneder T (2010) Die Agrarstrukturen in den Alpen und ihre Entwicklung unter Berücksichtigung ihrer Bestimmungsgründe: Eine alpenweite Untersuchung anhand von Gemeindedaten. Hebert Utz Verlag GmbH, München
Streifeneder T, Tappeiner U, Ruffini F, Tappeiner G, Hoffmann C (2007) Selected aspects of agro-structural change within the Alps. J Alp Res 95:41–52
Swiss Federal Statistical Office (2011) Landwirtschaftliche Betriebsstrukturerhebung 2011, Neuenburg
Tappeiner U, Cernusca A (1998) Effects of land-use changes in the Alps on exchange processes (CO_2, H_2O) in grassland ecosystems. In: Proceedings of the HeadWaler'98 conference held at Meran/Merano, Italy, April 1998, IAHS Publ. no. 248, pp 131–138
Tappeiner U, Tasser E, Leitinger G, Tappeiner G (2006) Landnutzung in den Alpen: historische Entwicklung und zukünftige Szenarien. In: Psenner R, Lackner R (eds) Die Alpen im Jahr 2020. Alpine Space—Man & Environment, vol 1. Innsbruck university press, Innsbruck, pp. 23–39
Tappeiner U, Borsdorf A, Tasser E (eds) (2008) Mapping the Alps. Spektrum, Heidelberg
Tasser E, Tappeiner U (2002) Impact of land use changes on mountain vegetation. Appl Veg Sci 5:173–184
Tasser E, Mader M, Tappeiner U (2003) Effects of land use in alpine grasslands on the probability of landslides. Basic Appl Ecol 4:271–280
Tasser E, Walde J, Tappeiner U, Teutsch A, Noggler W (2007) Land-use changes and natural reforestation in the Eastern Central Alps. Agric Ecosyst Environ 118:115–129
Tasser E, Leitinger G, Pecher C, Tappeiner U (2011) Agricultural changes in the European Alpine Bow and their impact on other policies. In: European Grassland Federation EGF: Proceedings of the Symposium of the 16th European Grassland Federation on "Grassland farming and land management systems in mountainous regions". Zürich: European Grassland Federation EGF (= Grassland Science in Europe 16), pp 27–38
Tasser E, Pecher C, Schreiner K, Tappeiner U (2012a) Wirtschaft schafft Landschaft. In: Tasser E, Schermer M, Siegl G, Tappeiner U (eds) Wir Landschaftmacher. Vom Sein und Werden der Kulturlandschaft in Nord-, Ost- und Südtirol. Verlagsanstalt Athesia, Bozen, pp 103–148
Tasser E, Schermer M, Siegl G, Tappeiner U (2012b) Wir Landschaftmacher. Vom Sein und Werden der Kulturlandschaft in Nord-, Ost- und Südtirol. Verlagsanstalt Athesia, Bozen
van Meijl H, van Rheenen T, Tabeau A, Eickhout B (2006) The impact of different policy environments on agricultural land use in Europe. Agric Ecosyst Environ 114:21–38
Verburg P, Overmars K (2009) Combining top-down and bottom-up dynamics in land use modeling: exploring the future of abandoned farmlands in Europe with the Dyna-CLUE model. Landsc Ecol 24:1167–1181
Verburg P, van Berkel D, van Doorn A, van Eupen M, van den Heiligenberg H (2010) Trajectories of land use change in Europe: a model-based exploration of rural futures. Landsc Ecol 25:217–232
Vittoz P, Randin C, Dutoit A, Bonnet F, Hegg O (2009) Low impact of climate change on subalpine grasslands in the Swiss Northern Alps. Glob Chang Biol 15:209–220
Weiss Sanpietro T, Weiss H, Hohl U (2004) Transfer-Struktur und Dynamik von Tranfers in die Testgebiete Safiental GR, Sernftal GL und Binntal VS. Universität St. Gallen, St. Gallen
Westhoek HJ, van den Berg M, Bakkes JA (2006) Scenario development to explore the future of Europe's rural areas. Agric Ecosyst Environ 114:7–20
Wilson GA (2007) Multifunctional Agriculture. A Transition Theory Perspective. CABI, Wallingford
Zimmermann P, Tasser E, Leitinger G, Tappeiner U (2010) Effects of land-use and land-cover pattern on landscape-scale biodiversity in the European Alps. Agric Ecosyst Environ 139:13–22

Food Security Challenges and Agricultural Development in Tajikistan

Kamiljon T. Akramov

1 Introduction

Evidence suggests that countries with mountainous terrain are more susceptible to food insecurity. It is estimated that about 40 % of the mountain population in developing countries (nearly 300 million people) is vulnerable to food insecurity, and nearly 90 % of this population lives in rural areas, almost half of which is likely to be chronically hungry (FAO 2002; Huddleston et al. 2003; von Dach et al. 2006). Past studies also show a significant gap in daily per capita calorie intake as well as per capita food availability between mountainous and non-mountainous countries. Limited availability of arable land resources constrains domestic agricultural production and makes mountainous countries heavily dependent on food imports. Further, high reliance on food imports makes such countries vulnerable to commodity price shocks, which creates additional food security challenges (Akramov et al. 2010).

Thus, it is not surprising that small import-dependent mountainous countries were severely affected by recent agricultural commodity prices shocks. High and volatile commodity prices clearly worsened the national- and household-level food security in these countries (Heady and Fan 2010; Akramov et al. 2010; FAO 2011a; Akramov and Shreedhar 2012). Agricultural commodity prices are more likely to remain on a high plateau throughout the next decade due to growing demand from consumers in emerging economies and supply-side challenges caused by increasing natural resource pressures and the global slowdown in projected yield improvements of important crops. Long-term climatic changes, the increasing frequency of unpredictable weather shocks; and growing linkages between agricultural commodity, energy, and financial markets may also amplify

K. T. Akramov (✉)
International Food Policy Research Institute, Washington, DC, USA
e-mail: k.akramov@cgiar.org

S. Mann (ed.), *The Future of Mountain Agriculture*, Springer Geography,
DOI: 10.1007/978-3-642-33584-6_9,

international food price volatility. Therefore, a combination of investment in improved agricultural productivity and increased food production will be needed to meet the rising demands for food and achieve sustainable long-term food security (FAO 2011a; OECD/FAO 2012). Increasing and achieving sustainable agricultural productivity is especially important for mountainous countries to increase domestic food production and mitigate potential negative impacts of future agricultural commodity price shocks.

After the recent turbulence in global agricultural commodity markets and its significant negative impact on food security, policymakers and development partners in Tajikistan, a small mountainous country located in Central Asia, are looking for ways to increase domestic food production in the country to moderate its reliance on food imports. Given the limited and declining availability of arable land resources, only substantial improvements in agricultural productivity and crop yields will enable mountainous countries such as Tajikistan to increase food production and ensure long-term food security. Advances in agricultural productivity, however, can be achieved by removing existing policy and institutional constraints and improving farmers' incentives.

The rest of the paper is organized as follows. Section 2 provides a brief overview of agricultural sector in Tajikistan. Section 3 discusses the impact of the recent agricultural commodity price shocks on food security and domestic food prices in the country. Section 4 discusses trends in agricultural productivity growth and constraints for agricultural development in Tajikistan. Section 5 concludes by summarizing the main findings of the study and outlining some policy recommendations.

2 Agriculture in Tajikistan

Tajikistan is home to about 8 million people, nearly 74 % of which resides in rural areas, and about 55 % is employed in the agricultural sector. Nearly 93 % of the country's land area is covered by mountains with high prevalence of steep lands (54 %), shallowness (48 %), and erosion risk (26 %). As a result, Tajikistan has limited arable land resources and only around 5.3 % of its land area is suitable for agricultural production (FAO 2000 and World Bank 2010). The country has the one of the lowest amounts of arable land per person (about 0.1 ha), which has been declining over the last two decades due to high population growth and land degradation (Akramov and Shreedhar 2012). Since Tajikistan has a highly mountainous terrain, most of its agricultural production is restricted to irrigated river valleys, which are broadly grouped into four valley systems as follows: Fergana Valley in the north along the Syrdarya river; Khatlon lowlands in the southwest; Gissar valley near Dushanbe; and narrow Zeravshan valley extending east to west between Fergana and Gissar valleys. There are four provinces, namely, Sughd (Fergana and Zeravshan valleys), Khatlon (Khatlon lowlands), Region of Republican Subordination (Gissar valley), and Gorno-Badakhshan

Autonomous Oblast (GBAO; Pamir highlands). Of these, Sughd and Khatlon are more populated, housing 32.9 and 39.3 % of the population with 17.8 and 17.3 % of the land area (24 and 33 % of agricultural land), respectively (Lerman and Sedik 2008).

Prior to the early 1990s, land in Tajikistan as elsewhere in the former Soviet Union was solely owned by state and agricultural production was dominated by large-scale socialist farms. The net effect of limited land reform since then was a significant shift in the distribution of agricultural land use among different farm types. The total amount of agricultural land allocated to state and collective farms declined significantly. As shown in Table 1, the land controlled by state and collective farms dropped dramatically, from more than 95 % of the total arable land in the pretransition period to only 27.5 % in 2009.[1] Most of agricultural land shifted to private dekhan farms and household plots in the process of limited land reform. In 2009, there were more than 18,000 private dekhan farms, with an average farmland size of 18 ha. They cultivated more than 40 % of the total (sown) arable land in the country. Nearly a quarter of the arable land was controlled by more than 740,000 traditional household plots that have an average size of 0.3 ha per holding. Another important feature of the changes in arable land use during this period was the allocation of so-called presidential plots to households, which now accounts for about 9 % of total arable land. There are about 375,000 such plots in the country with an average size of 0.2 ha per holding (FAO 2009).

Irrigated farming, which comprises more than two thirds of arable land in Tajikistan, is a major component of the country's agricultural sector. The irrigation network in Tajikistan was originally built to serve large-scale farms. The individualization of agricultural land use created an institutional vacuum, as no organization was responsible for the operation and maintenance of the on-farm irrigation networks. This problem was heightened by the collapse of public funding for the operation and maintenance of such networks. In order to address this problem, the Tajik government instituted an irrigation service fee that is to be paid by farmers to irrigation water providers, and used to operate and maintain the off-farm irrigation networks. Also, with the support from international development partners, the government is widely implementing collective action management solutions based on water users associations (Akramov and Shreedhar 2012).

The increased importance of food security led to reallocation of more arable land to food crops in recent years. As a result, the area sown with wheat, potatoes, and vegetables increased at the expense of cotton and feed crops.[2] Analysis of the composition of the sown area suggests that cereals are becoming the most

[1] Reallocation of agricultural land from public sector (state and collective farms) to private dekhan farms is continuing in Tajikistan. The data from the Agency on Statistics under President of the Republic of Tajikistan suggests that the share of public sector in total arable land declined to 18.4 % in 2010 (Statistical Agency, Statistical Agency (under the President of the Republic of Tajikistan) 2012; Accessed July 15, 2012. http://stat.tj/en/analytical-tables/real-sector/).

[2] The anecdotal evidences suggest that the area sown with cotton has slightly increased in 2011 due to favorable conditions in international cotton markets.

Table 1 Farm types and structure in Tajikistan

Farm type	Number	Average size (ha/farm)	Arable area (%)
State farms	193	322	7.5
Collective *dekhan* farms	9,000	18	20
Private *dekhan* farms	18,040	18	39.2
Household plots	740,400	0.3	24.2
Presidential plots	375,000	0.2	9.1

Source FAO 2009

Fig. 1 Distribution of agricultural land by crops, 2009. *Source* Statistical Agency (2012)

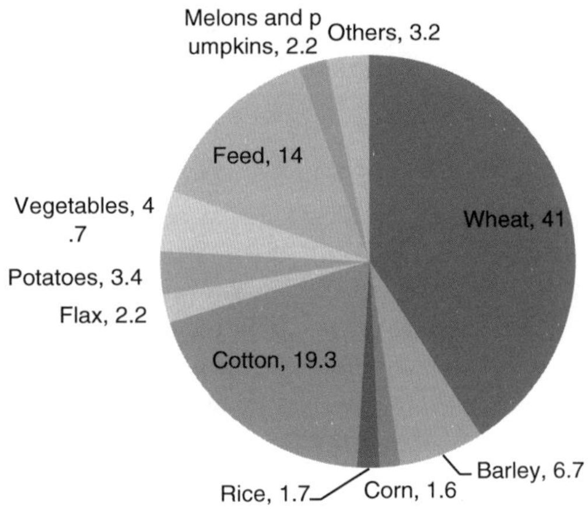

important crop in Tajikistan, accounting for about half of arable land in the country, with wheat being cultivated on 41 % of the land sown (Fig. 1). The second important crop by sown area is cotton, with almost 20 % of total sown area, which declined dramatically compared with the early 1990s. Potatoes, melons, fruits, and vegetables together account for about 10 % of total sown area. Area under these crops increased considerably during last several years. For example, land area under fruit and vegetable production increased by 16 and 30 %, respectively, compared with 2005 (Development Coordination Council 2011). On the other hand, the share of arable land allocated to feed crops declined significantly over the last decade, resulting in only 14 % of arable land currently being allocated to feed production (Akramov and Shreedhar 2012).

These changes in land allocation led to significant increases in grain, potato, and vegetable production, while production of cotton and feed crops has declined significantly (Fig. 2). The gross agricultural output in the grain sector, mainly wheat, has grown more than fourfold over the 1990–2009 period, with an average annual growth rate of 7.7 %. The annual growth rates of grain output have accelerated during the last decade (at more than 10 % annual average growth rate). Likewise, the potato output has tripled during this period, while growing with average annual growth rate of 11.2 %. The observed growth in grain and potato outputs is not due solely to changes in sown area. The grain yields also increased

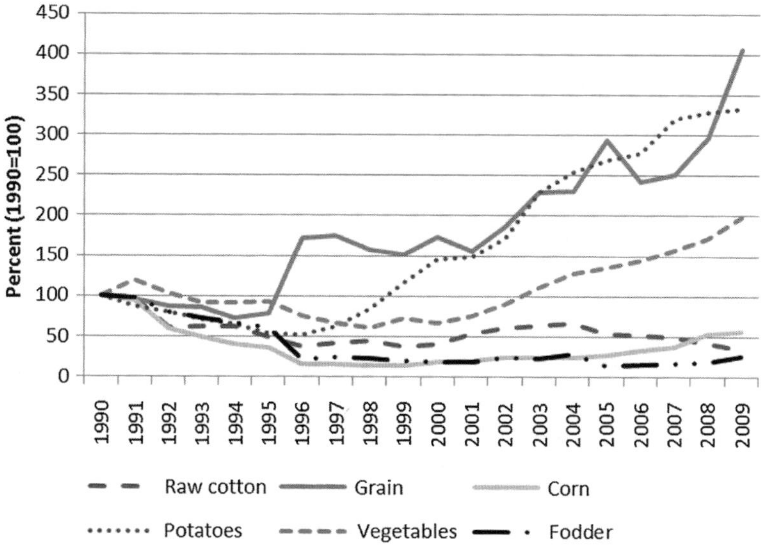

Fig. 2 Changes in gross crop output, 1990–2009. *Source* Statistical Agency (2012)

considerably during the last decade, probably due to increased allocation of irrigated land under wheat production. At the same time, the decline in feed crop cultivation combined with significant decreases in yields led to a shortage of fodder for the livestock sector. As a result, the livestock sector's share in gross agricultural output decreased from nearly 33 % in 1990 to less than 10 % in 2009 (Statistical Agency 2012).

3 Impact of Global Agricultural Commodity Prices on Food Security in Tajikistan

As mentioned earlier, Tajikistan heavily depends on food imports to meet its domestic food demands. Net imports account for more than 50 % of cereal consumption and more than 60 % of per capita calorie intake in the country comes from cereals. Moreover, nearly half of total meat consumption, including about 30 % of bovine beef and 80 % of poultry meat, also depends on imports. The heavy dependence on food imports means that Tajikistan is highly vulnerable to possible periods of food insecurity arising from external shocks such as turbulences in international commodity prices, food availability, and policy directives of its trade partners. In this regard, Tajikistan was among the hardest hit by the recent spike in agricultural commodity prices. The evidence suggests that global food and financial crises hit the country in multiple ways: economic growth contracted, the import prices for wheat and other staple food items soared, macro-level food security worsened and the crises negatively affected the welfare and food security of households (Akramov and Shreedhar 2012).

The ratio of food imports to total exports of goods and services, which reflects the relative cost of access to food available on the international markets, is often used to measure food security at the macro level (Diaz-Bonilla et al. 2002; Ecker et al. 2010). The lower this ratio, the more protected the country is in terms of macro-level food security. Based on this measure, from 2005 to 2008, on average, Tajikistan used more than 20 % of its export revenue to finance food imports, a figure more than two times higher than the world average.[3] Moreover, this ratio increased significantly in recent years, exposing Tajikistan's vulnerability to rising global food prices. From 2009 to 2011, about 35–40 % of Tajikistan's export revenues were spent to finance its food purchases in international markets (Akramov and Shreedhar 2012).

The present section now provides more formal empirical evidence on price transmission from international markets to domestic food prices in Tajikistan.[4] For these purposes the monthly time-series data for international wheat prices provided by the International Monetary Fund's (IMF) Primary Commodity Prices database are used[5] (IMF 2012). The focus on international wheat prices is justified by the fact that more than 60 % of daily calorie intake in Tajikistan comes from wheat and wheat products. Monthly domestic price series for wheat, wheat flour, bread, and aggregate food and consumer price indices were obtained from the Agency on Statistics of the Republic of Tajikistan. The price indices for tradable goods as well as the aggregate food and consumer price indices can all be affected by exchange rate movements. Thus, the data on monthly exchange rate movements was obtained from the National Bank of Tajikistan (National Bank of Tajikistan 2011). The X12-ARIMA procedure was used to remove the seasonal component from all price series.

Following Robles and Torero (2010), the first exercise to find evidence of price transmission was simply to analyze a graphical representation of the available price series. We computed 12-month growth rates for each price index, as well as the exchange rate, for every month since January 2000. The graphical analysis suggests that the growth rate of domestic wheat prices in Tajikistan closely follows the movement in international wheat prices (Fig. 3). This can be considered as a first basic indication of price transmission from international wheat markets to domestic wheat markets. This evidence is far from conclusive, however, as some food prices also experienced periods of high growth when international wheat prices were relatively stable, with rates that were well above the domestic inflation rate, especially for wheat flour. This suggests that other factors can affect domestic

[3] The average value of this indicator for 178 countries around the world is approximately 9 % (Ecker et al. 2010). Based on this measure, Tajikistan's macro-level food security is significantly lower than the all of its Central Asian neighbors.

[4] This discussion mainly draws from Akramov and Shreedhar (2012).

[5] This data set is available at www.imf.org/external/np/res/commod/index.asp.

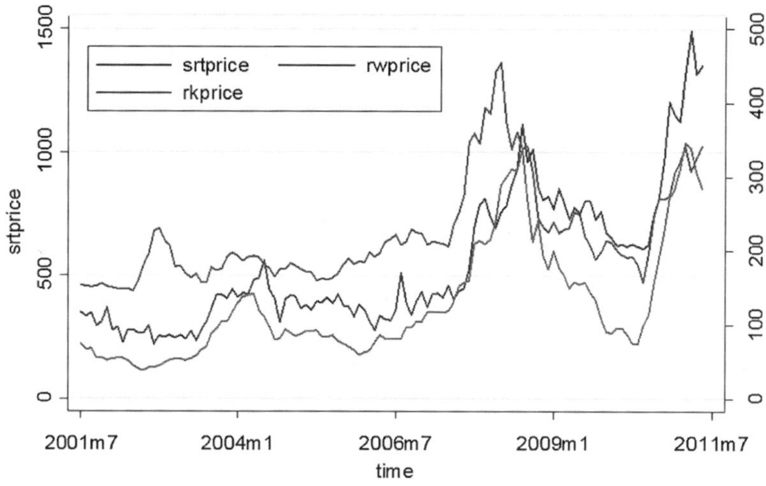

Fig. 3 Twelve-month growth rates of wheat prices in global markets, Kazakhstan and Tajikistan. *Source* Author's representation using data from IMF (2011), Kazakh-Zerno (2011), and Statistical Agency (2012). *Note* The data on wheat prices in Kazakhstan, Tajikistan's main wheat exporter, is included in the graph for comparison only. Our quantitative analysis mainly focuses on international wheat prices srtprice, rkprice, and rwprice are twelve month growth rates of wheat prices in Tajikistan, Kazakhstan and international markets, respectively

food prices even when international prices are stable. Nevertheless, acceleration of domestic prices after 2007 seems to be quite general or at least more obvious than before 2007. Graphical representation also suggests that international wheat prices may affect not only domestic prices of wheat products but also overall food and consumer price inflation rates (These graphs are not provided here).

Further, augmented Dickey–Fuller and Phillips–Perron tests with and without a time trend and with no price lags, one lag, two lags, three lags, and so forth up to seven lags suggest that domestic wheat market in Tajikistan is cointegrated to international wheat markets with order of one, which is as expected. Therefore, instead of running regressions at levels, the Johansen cointegration test was used. The underlying vector autoregression (VAR) model in this test included domestic and international wheat prices and the exchange rate. The results show strong evidence of cointegration relationships between changes in international wheat prices and domestic food prices in Tajikistan.

Finally, moving-average first-difference regression models were estimated to test whether the growth rate of international prices had explanatory power on the growth rate of domestic food prices. In these regressions, the growth rate of international wheat prices and the depreciation rate of domestic currency, somoni, are included as explanatory variables. The fixed effects were also included to control for the period between mid-2007 and the end of 2008, when international wheat prices were very high. The results are reported in Table 2. These results show that the growth rate of international wheat prices positively affects the

Table 2 Evidence on transmission of international wheat prices to domestic food prices in Tajikistan

	Wheat	Wheat flour	Bread	Food price	CPI
Global wheat prices (GWP)					
One-month lag	0.456	0.058	0.029	0.003	–0.000
	(0.158)**	(0.052)	(0.036)	(0.018)	(0.013)
Two-month lag	0.038	0.169	0.035	0.046	0.028
	(0.158)	(0.053)**	(0.035)	(0.018)**	(0.013)**
Global food crisis (GFC)	0.601	0.698	–0.123	0.309	0.147
	(2.289)	(0.764)	(0.515)	(0.227)	(0.189)
Interaction of GWP and GFC	0.685	0.385	0.224	0.077	0.054
	(0.322)**	(0.105)**	(0.071)*	(0.037)**	(0.026)**
Exchange rate	–0.240	–0.488	–0.307	0.045	0.049
	(0.891)	(0.290)	(0.196)	(0.101)	(0.076)
R = squared	0.17	0.37	0.25	0.28	0.25
N	106	106	106	106	106
F-statistic	1.81	5.06	2.78	3.29	2.87

Source Authors' own estimations
Note CPI Consumer Price Index. * and ** significant at the 1 and 5 % levels, respectively

growth rate of the domestic prices of wheat and wheat products in Tajikistan. For example, during noncrisis periods, 1 % increase in international wheat prices causes about 0.5 % increase in domestic wheat prices (with one-month lag) in Tajikistan. However, during a global food crisis period, 1 % increase in international wheat prices causes more than 1.1 % increase in domestic wheat prices. In addition, the results suggest that fluctuations in international wheat prices not only have significant causal impact on domestic wheat and wheat product prices in Tajikistan, but the overall food and consumer price inflation rates in the country are also noticeably affected. For example, 1 % increase in international wheat prices is associated with 0.08 % increase in aggregate consumer price inflation. Overall, our results suggest that that there is strong empirical evidence of price transmission from international wheat markets to the domestic food markets in Tajikistan.

In this situation, a change in international wheat prices affects producers and consumers in Tajikistan in opposing direction: consumers lose and producers gain from an increase in wheat prices, and vice versa when prices decline. In fact, the evidence suggests that recent spike in agricultural commodity prices had significant negative impact on household welfare in Tajikistan. More than 50 % of households in Tajikistan reported reducing staple food consumption as a result of the global crisis, as compared with 38 % in all transition countries and 35 % in other Central Asian countries (EBRD, 2011; Akramov and Shreedhar 2012). Another implication of rising agricultural commodity prices is that farmers will gain additional incentives to increase food production in the country. However, these additional incentives may not be straightforward, as these incentives will be affected by natural resource constraints, and various policy and institutional factors present in the country. The subsequent section will discuss such challenges being important for the future of agricultural production and food security in Tajikistan.

4 Constraints on Agricultural Development in Tajikistan

The evidence presented above suggest that high reliance on international markets makes Tajikistan vulnerable to changes in global agricultural commodity prices. It is widely expected that agricultural commodity prices will remain at high levels throughout the next decade, supported by growing demand for food and fuel (FAO 2011a; OECD/FAO 2012). This perspective will create additional food security challenges for Tajikistan given its high reliance on imported food, especially wheat and wheat products. Moreover, demand for food in the country is expected to increase due to rising population and per capita income.[6] Tajikistan needs to increase domestic food production to cope with growing demand for food and high agricultural commodity prices in international markets. Given Tajikistan's terrain and agro-ecological conditions and technology, the feasible scope for expansion of arable land area for crop production is very limited. That is why increasing crop yields and agricultural productivity is critical to meet growing demand for food and to decrease the country's reliance on food imports, and as such, it is instructive to examine recent performance and constraints to increasing crop yields and agricultural productivity in the country.

4.1 Trends in Agricultural Productivity

The partial productivity of land, measured as aggregate value of agricultural value added per hectare of arable land, declined nearly two times between 1991 and 1996 (Fig. 4). Agricultural value added increased significantly since 1997, while arable land has remained generally constant. This led to a dramatic increase (in constant 2000 prices) in land productivity between 1997 and 2010, which almost tripled. Agricultural labor, unlike agricultural land, showed steady increase in Tajikistan since early 1990s. The collapse of agricultural production in the early 1990s led to a dramatic decline (nearly three times) in partial productivity of agricultural labor, measured as aggregate value of agricultural value added per worker, between 1991 and 1997. The growth in agricultural value added over performed the growth in agricultural labor after 1997 and as a result the partial productivity of labor also increased between 1997 and 2010 (Fig. 4). However,

[6] Data from the National Statistical Agency of Tajikistan suggest that the average population growth rate during the last decade has been about 2 % per year. According to UN population projections, Tajikistan's population is expected to grow at an annual rate of 1.5 % throughout next two decades and may reach 9.5 million by 2030.

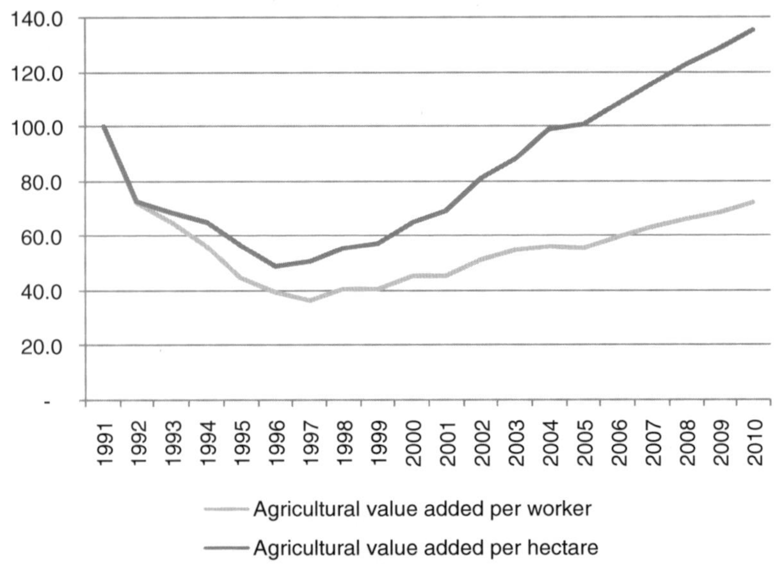

Fig. 4 Agricultural productivity in Tajikistan, 1991–2010. *Source* Author's computations using data from World Bank (2011)

despite such trend agricultural labor productivity in Tajikistan still significantly lower than in early 1990s. Further, the evidence also suggests that growth in agricultural total factor productivity resumed after the collapse in the early 1990s (Fuglie 2012). Nevertheless, agricultural productivity growth is still low in Tajikistan. As a result, current levels (2010) of the partial productivity of land and labor in Tajikistan's agriculture remain very low at US$1238 and US$736, respectively.

The growth in agricultural productivity inevitably depends on the evolution of crop yields. Despite some growth, realized crop yields in Tajikistan are still well below both average yields observed in neighboring countries and what may be considered economically attainable yields in the context of prices and costs. For example, wheat yields of around 2 tons per ha in Tajikistan are significantly lower than average yields observed in other Central Asian countries (Fig. 5). This is especially worrying given the evidence that crop yield gaps[7] in Central Asia are one of the greatest among geographic regions of the world. The crop yield gap for Central Asia is estimated at 64 %. This means that the potential to increase output is particularly high in the region, including Tajikistan. Since farming in Tajikistan is running far below from potential, reducing gaps in crop yields would mean significant increase in output, which can have significant implication for food security in the country.

[7] Yield gaps are estimated by the differences between economically viable yield potential and average farmers' yields in a given country over some specified temporal scale of interest.

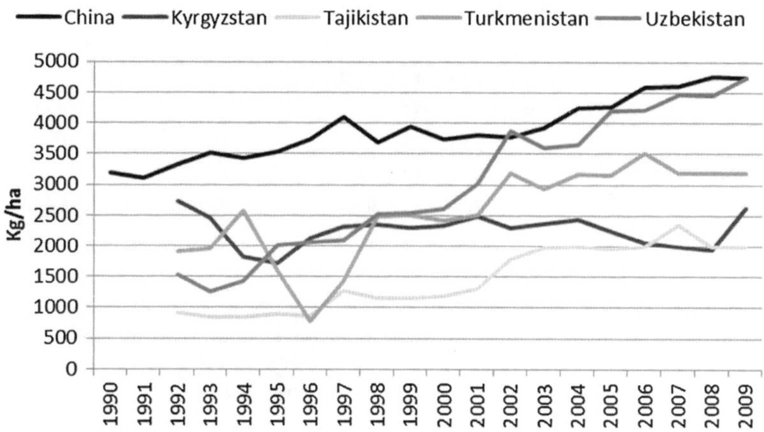

Fig. 5 Wheat yields in selected countries, 1990–2009. *Source* FAO (2011c)

The important question is that what can be done in order to increase crop yields and agricultural productivity in Tajikistan. There are two components of yield gaps (FAO 2011b; OECD/FAO 2012). The first component is associated with agro-ecological and environmental conditions and it basically cannot be reduced. The second component is associated with crop management practices and institutional and policy factors. This second component can be reduced, which can help to improve agricultural productivity. The potential for improvement of crop yields and agricultural productivity in former socialist countries like Tajikistan is dependent on various factors including existence of secure property rights, development of market infrastructure and collective action institutions, and availability and affordability of appropriate technologies (Rozelle and Swinnen 2004). These factors are particularly important for the future of agriculture and food security in Tajikistan, where additional difficulties were created by the civil war that raged in this country until 1997.

4.2 Constraints to Agricultural Productivity Growth

There is a certain consensus among economists that secure property rights lead to better economic outcomes. The literature considers three channels for this positive link (Do and Iyer 2008). First, stable property rights provide greater incentives for labor efforts and long-term investments in land and new agricultural technology. A second channel is through the enhanced possibilities for transfer of land to more efficient users. Third, if better property rights make it easier to use land as collateral, then agricultural growth might be improved through better access to credit and reduced constraints on investment. In this regard, agricultural land in Tajikistan is solely owned by the state, and private property rights are limited to use rights. The initial steps of land reform were in the early and mid-1990s, when additional land was allocated to households and an attempt was made to reorganize larger state farms into corporate farms (Lerman and Sedik 2008). Since the end of civil war in the mid-1990s, the state land reforms became an important policy goal because these large farms were characterized by low productivity and increasing inefficiency. In collaboration with development partners, the government made several attempts to develop appropriate farm restructuring and land reform strategies. The initial reforms attempted to restructure the unprofitable and marginally profitable collective and state farms into associations of leaseholders and dekhan farms and to transform state farms into collective farms. At the same time, Presidential Decrees 342 and 874 (1995 and 1997, respectively) allocated 75,000 ha of irrigated land to households for private farming. From 1999 to 2005, more than 500 large collective farms were restructured into smaller collective dekhan farms. In 2006, about 200 additional large collective farms were restructured into smaller farms. Overall, during this period, a total of nearly 700 large collective farms were restructured into more than 5,000 individual dekhan farms. In general, official statistics suggest that an estimated 45 % of total arable land was affected by farm restructuring and land reform by 2007 (Development Coordination Council 2011).

However, the evidence suggests that the productivity of private farms has not increased significantly and is still significantly below than productivity levels on household gardens. Lack of significant improvement in property rights and delays in issuing land certificates, in combination with the absence of land cadasters, are limiting farmers' ability to make appropriate and efficient production decisions (Lerman and Sedik 2008; World Bank 2010). Moreover, it is important to note that land reform and farm restructuring are incomplete, and reallocation of agricultural land from public sector (state and collective farms) to private dekhan farms is still continuing in Tajikistan. The more recent data from the Agency on Statistics under President of the Republic of Tajikistan suggests that as of 2010 about 18 % of arable land still remained in public (collective) sector (Statistical Agency 2012). Thus, completion of land reform and creation of secure property rights are vital in increasing agricultural productivity in the country. It would be useful to better understand and assess the current status of farming systems, their efficiency, and potential institutional and policy reforms—while considering lessons from other transition countries such as Armenia, Georgia, Kyrgyzstan, Moldova and China—to take further concrete steps in the near future.

Another major problem in Tajikistan's agricultural sector is related to insufficient collective action in the delivery of and access to irrigation water, which has been drastically affected by individualization of land use. The government created a legal framework to reform the irrigation sector through establishment of water users associations (WUAs). However, merely changing a regime and creating a legal framework seem insufficient for successful institutional change in irrigation. The evidence suggests that important societal features inherited from socialism and amplified during the transition process, such as incongruities between formal and informal rules, power abuse, rent seeking, information asymmetry, and distrust between community members, played a crucial role in evolution of social capital and collective action in the irrigation sectors of other transition countries (Theesfeld 2004; Akramov and Omuraliev 2009). The volume of anecdotal evidences suggests that similar problems related to collective action are applicable to Tajikistan's irrigation sector.

As a result, the irrigation infrastructure is inadequately funded due to low collection of water user fees and the majority of this infrastructure is in deteriorating condition. The existing evidence highlights this fact: the efficiency of irrigation systems is estimated to be around only 55 %, and nearly 70 % of the boreholes for vertical drainage are estimated to be out of order. More than 50 % of the gravity irrigation systems and about 65 % of pumping stations are estimated to have deteriorated. Despite a wealth of water resources, around 20 % of the irrigated lands such as those in Kulab, Istravshan, and Gissar regions face water shortage due to poor regulation of river flows (MIWMRT, UNDP, and IFSAS 2006; Development Coordination Council 2011). It is important to note that the evidence suggests that investments in irrigation infrastructure to reduce the effects of weather on agricultural production can be an important means to improve agricultural productivity and ensure stable agricultural performance (Fan et al. 2008).

Next impediment to increase agricultural productivity and crop yields in Tajikistan is poor supply of agricultural inputs. The dissolution of the previous public agriculture support system has left the sector with a poor input supply and inadequate agricultural extension support system. In order to increase crop yields and agricultural productivity, the availability and affordability of modern inputs such as improved seeds, fertilizer, farm machinery, and agricultural advisory services and access to credit are necessary. In the previous socialist system, the state provided the former state and collective farms with improved seeds, fertilizer and other inputs for farming and provided farm managers, agronomists, and technical staff to ensure modern farming practices. In the process of ongoing land reform and liberalization of agricultural markets, the old system has been abolished but the new market-based system has not been created yet (Lerman and Sedik 2008; World Bank 2010; Development Coordination Council 2011).

Now improved seeds, fertilizer and other farm inputs are mostly sold by private actors, although in some cases cotton seed prices are sometimes controlled by ginneries (ICARDA 2008). The seed demand exceeds the supply, and only about 50 % of improved crop seeds demanded are supplied now. As a result, farm-saved seeds constitute the main source of seeds, and unchecked and uncertified seed sale and use have been reported (FAO 2009). Poor seed storage facilities are also a constraint. The utilization of improved seeds has been instrumental in improving agricultural productivity and food security in many developing countries, as seen from the experiences with hybrid rice in China and the Green Revolution in Asia (Li et al. 2009; Hazell 2009). Apart from raising investment in domestic seed research capacity, collaboration with international institutes and the private sector may help to improve and increase quality seed available for farmers. In addition, fertilizer use has also declined considerably and is estimated to average somewhere between 26 and 30 kg/ha. Past studies observed that a considerable monopoly exists in fertilizer supply, as large cotton companies' supply up to 75 % of the fertilizer available in the market (World Bank 2010).

Further, Lerman and Sedik (2008) observed that machinery inventories in Tajikistan have declined dramatically from the Soviet period. The number of tractors declined from 37,000 in 1991 to 19,000 in 2006, grain harvesters from 1,500 to 1,600 in the early 1990s to 900 in 2006, and cotton harvesters from 3,000 in 1991 to fewer than 600 in 2006. Field reports also reveal that the quality of tractors has deteriorated considerably because of significant costs associated with replacing parts. Smaller farms rely heavily on hiring tractors at increasing variable costs. While newer tractors are increasingly available through the market, given the considerable cost associated with purchase, usage still seems to be declining.

It is important to note that there is a concerted public effort to improve agricultural input delivery in the country. For example, in the seeds sector, laws on crop-breeding achievements, plant quarantine, and the seed industry were adopted in 1995, 2001, and 2008, respectively, to deal with various aspects of seed breeding, regulation of variety testing, registration, seed quality control, certification, and other issues. The current reform agenda envisions that the supply of inputs, including seeds, fertilizers, and pesticides will be channeled via the market, and farmers' service cooperatives are envisioned to play an important role in improving farmer access to such inputs (Development Coordination Council 2011). Along with input supply, these service cooperatives are expected to strengthen appropriate farming practices, agronomic best practices, and methods of input use, with help from the private sector, research, and donor agencies. These agricultural services are currently dispensed through a technical staff at the local government which is accountable to both relevant ministries and local government officials. International experience suggests that smaller farmers can make use of small-scale economies and pool resources together to increase access to inputs and services (Lerman et al. 2004). In this regard, development partners (such as the German Agency for International Cooperation, or GIZ, and International Finance Cooperation or IFC) are collaborating with the government to develop agricultural machinery leasing services and trade and service cooperatives in Tajikistan. However, the lack of awareness and institutional support for building capacity among these cooperatives and the ability to link them to potential partners are major constraints in delivery of efficient services.

A practical agricultural finance model piloted by GIZ, the European Union (EU), and other development partners under the Tajik Agricultural Finance Framework (TAFF) project is a value-chains approach. This approach attempts to streamline agricultural support systems from input supply and extension service delivery to output marketing, transportation, storage, processing, packaging, and branding by developing the private-sector participation. This can be a useful approach to develop and modernize the agricultural sector in the country. One of the components is fee-based technical assistance for farmers through field advisors who patrol the crop fields, check production status, and report back to a main agronomist, who then dispenses the advice. Currently there are 13 technical assistance groups with 60 field advisors that cover 14,000 ha of land, mainly in

Khatlon and Sughd provinces. Their focus is primarily on cotton, with scope to scale up such services to other crops. Farmers currently pay 50 % of the cost (total cost is 50 somoni (or about $11.42) per ha), which is what the farmers are currently willing to pay as cotton prices are on the rise. The tentative reception to this scheme seems positive, and advised farmers seem to perform better than those not advised. If successful, as determined by visibly increasing yields, such pilot programs can be tested and scaled up as agriculture develops into a feasible private-business venture (EBRD 2010; Development Coordination Council 2011).

Another major constraint to enhancing productivity and profitability of agriculture is the taxation of farmland, inputs, and outputs, combined with limited access to credit. For instance, farmers pay import taxes on inputs including 5 % import tax for input supplies, 20 % value added tax, plus customs fees that directly impact farmer incomes and therefore ability to invest back in the farm. Agricultural export taxes for cotton are around 10, and a 25 % social tax is also charged on farms (IMF 2011). Apart from streamlining agricultural taxation, new financing mechanisms for agriculture activities are also needed. The European Bank for Reconstruction and Development (EBRD)-TAFF project has been active since 2010 in devising and testing alternative financing practices through community-based and local microfinance initiatives, credit-savings schemes, savings groups, and microfinance institutions (Development Coordination Council 2011). Other financial alternatives to safeguard production such as crop insurance through public–private partnerships are at a discussion stage.

5 Conclusions

The evidence presented in this study shows that food security in Tajikistan is vulnerable to volatility of international agricultural commodity prices. The recent spike in international food prices deteriorated macro level food security and adversely affected household welfare in the country. The empirical results reported in this study suggest that there is strong evidence of price transmission from international wheat markets to domestic food markets in Tajikistan. The rising international wheat prices led to higher domestic food prices in Tajikistan, especially wheat and wheat product prices. This has important implications for the future of the agricultural sector in Tajikistan because it needs to increase domestic food production to reduce high reliance on food imports. One consequence of rising agricultural commodity prices is that farmers will gain additional incentives to increase domestic food production. However, these additional incentives may not be straightforward, as these incentives will be affected by natural resource constraints, and various policy and institutional factors present in the country.

The agricultural sector in Tajikistan has been constrained by limited arable land resources. Moreover, arable land resources are steadily declining in per capita terms due to land degradation and high population growth. Obviously, agricultural productivity growth is the only feasible way to improve long-term agricultural productivity and food security in the country. But the collapse of agricultural production in the early 1990s had significant negative impact on agricultural productivity. Despite decent growth rates in recent years, agricultural productivity in Tajikistan still remains significantly below than in the early 1990s. Further, crop yields in Tajikistan are significantly below the potential and there is significant yield gap. This suggests that the potential to increase food production is significant.

However, multiple institutional and policy constraints exist to agricultural production in the country. First, as of 2010, nearly one-fifth of arable land is still in public sector (collective farms) use. Since the productivity of private dekhan farms and household plots is significantly higher than the remaining collective farms, reallocation of this land to private dekhan farms is likely to increase overall agricultural output. In addition, significant improvements in the private property rights of farmers, such as giving farmers freedom to make appropriate production decisions based on market and agro-ecological conditions, are necessary to further increase crop yields and agricultural productivity.

Second, besides securing farmers' private property rights, it is imperative to promote and empower collective action institutions such as water users associations, producer organizations, and trade and service cooperatives. WUAs can help farmers to properly manage and maintain on-farm irrigation systems, effectively coordinate irrigation water use, and impartially allocate irrigation water among users. Trade cooperatives can help farmers' access to agricultural and consumer markets. Agricultural service cooperatives can help to improve farmers' access to modern inputs and services. Moreover, such institutions can develop mechanisms for resolving farmers' common economic problems that prevent them from becoming more productive and profitable.

Third, there is very little evidence in the literature on institutional and structural bottlenecks in rural Tajikistan that constrain agricultural productivity in Tajikistan. In the future, research that explores such constraints and further illuminates the role of institutional innovations in advancing agricultural productivity could potentially help the policy debate in the country. The findings of such research can also be helpful in formulating and implementing an effective agricultural development strategy in the country.

References

Akramov K, Omuraliev N (2009) Institutional change, rural services and agricultural performance in Kyrgyzstan. Discussion paper 904. International Food Policy Research Institute, Washington, DC

Akramov K, Shreedhar G (2012) Economic development, external shocks and food security in Tajikistan. IFPRI discussion paper 1163. International Food Policy Research Institute, Washington, DC

Akramov KT, Yu B, Fan S (2010) Mountains, global food prices, and food security in the developing world. IFPRI discussion paper 989. International Food Policy Research Institute, Washington, DC

Development Coordination Council (2011) International conference on agrarian reform. In: Conference proceedings. Government of the Republic of Tajikistan and Development Coordination Council, Dushanbe

Diaz-Bonilla E, Thomas M, Robinson S (2002) On boxes, contents, and Users: food security and the WTO negotiations. Trade and macroeconomics division discussion paper 82. International Food Policy Research Institute, Washington D.C

Do QT, Iyer L (2008) Land titling and rural transition in Vietnam. Econ Dev Cult Chang 56(3):531–580

Ecker O, Breisinger C, McCool C, Diao X, Funes J, You L (2010) Assessing food security in Yemen: an innovative integrated, cross-sector, and multilevel approach. IFPRI discussion paper 00983. International Food Policy Research Institute, Washington, DC

EBRD (European Bank for Reconstruction and Development) (2010) Tajik agricultural finance facility. Operation performance evaluation review. European Bank for Reconstruction and Developement, London

EBRD (European Bank for Reconstruction and Development) (2011) Life in transition: after the crisis. European Bank for Reconstruction and Developement, London

Fan S, Gulati A, Thorat S (2008) Investment, subsidies and pro-poor growth in India. Agric Econ 39:163–170

FAO (Food and Agriculture Organization of the United Nations) (2000) Land resource potential and constraints and regional and country level. Food and Agriculture Organization, Rome

FAO (Food and Agriculture Organization of the United Nations) (2002) The state of food security in the world 2002. Food and Agriculture Organization, Rome

FAO (Food and Agriculture Organization of the United Nations) (2009) FAO crop and food security assessment mission to Tajikistan. Mission report. Food and Agriculture Organization, Rome

FAO (Food and Agriculture Organization of the United Nations) (2011a) The state of food insecurity in the world 2011: how does international price volatility affect domestic economies and food security. Food and Agriculture Organization, Rome

FAO (Food and Agriculture Organization of the United Nations) (2011b) The state of the world's land and water resources for food and agriculture. Managing systems at risk. FAO conference document c2011/32. Rome, 25 June–2 July

FAO (Food and Agriculture Organization of the United Nations) (2011c) FAOSTAT statistical database. http://www.fao.org/economic/ess/ess-fs/en/. Accessed 15 Dec 2011

Fuglie KO (2012) Productivity growth and technology capital in the global agricultural economy. In: Fuglie KO, Wang SL, Ball VE (eds) Productivity growth in agriculture: an international perspective. CAB International, Oxfordshire

Hazell P (2009) Transforming agriculture: the green revolution in Asia. In: Spielman DJ, Pandya-Lorch R (eds) Millions fed: proven success in agricultural development. International Food Policy Research Institute, Washington, DC, pp 77–82

Heady D, Fan S (2010) Reflections on the global food crisis: how did it happen? how has it hurt? and how we can prevent the next one? International Food Policy Research Institute, Washington, D.C

Huddleston B, Ataman E, de Salvo P, Zanetti M, Bloise M, Bel J, Francheschini G, Fe' d'Ostiani L (2003) Towards a GIS-based analysis of mountain environments and populations. Working paper no. 10. Food and Agriculture Organization, Rome

ICARDA (International Center for Agricultural Research in the Dry Areas) (2008) Focus on seed programs: the Tajik Seed Industry. WANA seed network focus on seed programs country report. International Center for Agriculture in Dry Areas, Allepo. http://www.icarda.org/seed_unit/Pdf/Focus/FOCUS-Tajikistan.pdf

IMF (2012) Primary commodity prices database. http://www.imf.org/external/np/res/commod/index.asp. Accessed 15 March 2012

IMF (International Monetary Fund) (2011) Primary commodity prices database. http://www.imf.org/external/np/res/commod/index.aspx. Accessed 15 Dec 2011

Kazakh-Zerno (Information Agency Kazakh-Zerno) (2011) Statistical database on cereal prices. Petropavlovsk. Kazakhstan. http://kazakh-zerno.kz/. Accessed 15 Dec 2011
Lerman Z, Csaki C, Feder G (2004) Agriculture in transition: land policies and evolving farm structures in Post-Soviet countries. Lexington Books, Oxford
Lerman Z, Sedik D (2008) The economic effects of land reform in Tajikistan. Policy studies on rural transition No. 2008-1. FAO Regional Office for Europe and Central Asia, Budapest, Hungary
Li J, Xin Y, Yuan L (2009) Pushing the yield frontier: hybrid rice in China. In: Spielman DJ, Pandya-Lorch R (eds) Millions fed: proven success in agricultural development. International Food Policy Research Institute, Washington, DC, pp 77–82
MIWMRT (Ministry of Irrigation and Water Management of Republic of Tajikistan), UNDP (United Nations Development Program), and IFSAS (International Fund for Saving the Aral Sea) (2006) Water sector development strategy in Tajikistan. UNDP Tajikistan Country Office, Dushanbe
National Bank of Tajikistan (2011) Statistical database. http://nbt.tj/en/payments_balance/. Accessed 15 Dec 2011
OECD (Organization for Economic Cooperation and Development)/FAO (2012) OECD-FAO agricultural outlook 2012–2021. OECD Publishing, Paris. http://dx.doi.org/10.1787/agr_outlook-2012-en
Robles M, Torero M (2010) Understanding the impact of high food prices in Latin America. Economia 10(2):117–164
Rozelle S, Swinnen JJ (2004) Success and failure of reform: insights from the transititon of agriculture. J Econ Lit 42(2):404–456
Statistical Agency (under the President of the Republic of Tajikistan) (2012) Database of the statistical agency of the President of the Republic of Tajikistan. http://www.stat.tj/english/home.htm. Accessed 15 June 2012
Theesfeld I (2004) Constraints on collective action in a transitional economy: the case of Bulgaria's irrigation sector. World Dev 32(2):251–271
von Dach SW, Ott C, Klaey A, Stillhardt B (2006) Will international pursuit of the millenium development goals alleviate poverty in mountains? Mt Res Dev 26(1):4–8
World Bank (2010) Agricultural sector growth in Tajikistan: opportunities and constraints. Background paper for World Bank Country Economic Memorandum. World Bank, Washington, D.C
World Bank (2011) World development indicators database. http://data.worldbank.org/data-catalog/world-development-indicators. Accessed 15 Dec 2011

Impact of Natural Resources and Infrastructure on Future Livelihood in Mountain Farming in the Himalaya Region

Werner Doppler and K. C. Krishna Bahadur

1 Introduction

In many mountainous regions, deforestation, land degradation, unregulated water flows and subsistence orientation of farming are major development challenges to improve sustainable livelihood of mountain farming families. There are many reasons for these developments. The most prominent once are increasing population density, inadequate use of natural resources in the process of increasing scarcity of resources and the lacking of proper infrastructure and efficient transportation systems to connect people to markets, educational centers, health institutions and modern knowledge systems (Doppler et al. 2006).

In the Himalaya region this development is very much pronounced and influenced the development of farming systems and living standard of mountain farmers. Farming systems in the Himalayan region differ from other mountain zones such as Southeast Asia in the following ways:

1. The agricultural economy and market infrastructure in these states is less developed as in other mountain regions and hence the driving force for development in remote areas is more limited. Export of agricultural products occurs less than in countries like Thailand (e.g. rice) or Vietnam (e.g. coffee, tea).
2. General infrastructure is less developed partly due to very long distances from remote areas to urban centers and due to high altitudes conditions.

W. Doppler (✉)
University of Hohenheim, Stuttgart, Germany
e-mail: Werner.Doppler@uni-hohenheim.de

K. C. Krishna Bahadur
University of Guelph, Guelph, ON, Canada

S. Mann (ed.), *The Future of Mountain Agriculture*, Springer Geography,
DOI: 10.1007/978-3-642-33584-6_10,

3. Conditions for agricultural production differ especially in very high altitudes due to climatic conditions and are much determined by tradition, culture and social differentiations leading to a dominant role of subsistence farming.

In the Himalayan countries Nepal, Bhutan, Northern India and Northern Pakistan and in the regions of Tibet and Jammu, Kashmir and Ladakh farming systems are different according to changing natural and political environments. This applies especially to the degree of market liberalization. Differences regarding resource ownership such as land titles and water rights as well as for product marketing and exporting especially are obvious when comparing Nepal and Bhutan. There are complex relations between infrastructure and market distances, between traditional and modern knowledge and between natural and socio-cultural systems. Different social organizations such as the caste structures have an additional impact on the development of agricultural societies.

To assess rural resources and future livelihood development in mountain's farming system in Himalaya region a short view on the farming systems in the region is given. In Chap. 3 an area in Nepal is selected which can be considered as typical for many situations in the Himalayan region and problems relevant for similar areas will be investigated and the future potential discussed. Together with results of other research studies in the region conclusions are drawn for the future regarding the impact of natural resources and infrastructure on future livelihood in mountain farming in the Himalaya region in Chap. 4.

2 Farming Systems in the Himalayan Region

Changes in land use in the Himalayan region have been drastic in the last three decades. Forest land was reduced by more than 35 % in the last 30 years and agricultural land, especially irrigated land, increased. Today, paddy fields cover about 25 % of the area. In smaller mountainous sub-regions, this process was even more drastic.

Farming in the mountains of Himalaya region is strongly determined by subsistence and the transitions to more market orientation. Families live in a traditional environment and base their decisions on own resources and traditionally known techniques with a long-term survival strategy. Very traditional systems are sustainable, but on a very low economic level. In the course of time and with increasing population more pressure rests on the availability and use of natural resources. The more the families are related to markets and the more economic paradigms dominate, the tendency to use resources more intensively are obvious. In areas where the connection to markets and other facilities in urban areas improves a clear trend to higher living standard can be seen. This is in many cases related to more risk in farming and resource conservation. Not only inadequate use of land but also improper water management in large mountainous areas creates the danger of water erosion and land sliding. In more intensive land use areas, the use of external inputs is common, but not always under proper control as e.g. in the use of chemicals.

In the course of development and the increasing intensity of resource use the degree of scarcity of resources is increasing and ownership becomes an increasingly important issue. Unclear land ownership rights and water rights are more relevant the more investment into resource quality is needed and the more external capital is involved. The level of dependency on resource owners (land, water, capital) increases with increasing land scarcity and connectivity to markets. The transfer from subsistence to more market oriented farming systems has its price in the dependency on land lords, water lords and money lenders.

In the spatial context these relations can be seen alongside a gradient from urban to remote areas in a typical valley in Himalaya region (see Fig. 1) following the concept of Von Thünen (1826). Such a rural gradient analyses shows main features from urban centers to the most remote area of a mountain region as follows (Doppler et al. 2006, p. 27):

- Decreasing population density and decreasing size of settlements and hence decreasing size of social groups
- Increasing share of forests and at the highest altitudes no forests but alpine grazing areas
- Decreasing land scarcity often with ecological balanced farming (traditional way of protecting nature)
- Less clear regulations of ownership rights on resources
- Increasing market distance and decreasing opportunities for transportation
- Increasing share of subsistence-based families and decreasing connections to the monetary system
- Close to urban centers intensive use of agricultural lands with modern inputs and mechanization to very traditional production methods with local and natural inputs only in very remote areas and in between these zones very often the most dangerous development of inadequate use of resources
- High level of food security in very remote areas but on a low supply level, and low food security level an middle distance zones from urban and very remote areas
- Increasing traditional health care at low level and decreasing modern health care
- Decreasing educational opportunity and high level of traditional knowledge
- Changing ethnic or religious groups
- Increasing family ties and traditional cultures
- Decreasing man-made problems
- Deceasing income and living standards but increasing support amongst people.

Overall, income, administrative capacity, education, health care, market relations, ownership rights are decreasing while social, religious and cultural issues are more pronounced.

Farming systems development and the future potential towards more market orientation in Himalaya region has been dealt with in many research projects. For Nepal Karki (2004), Bahadur (2005), Goshi (2005), Das (2010), Maharjan (2010),

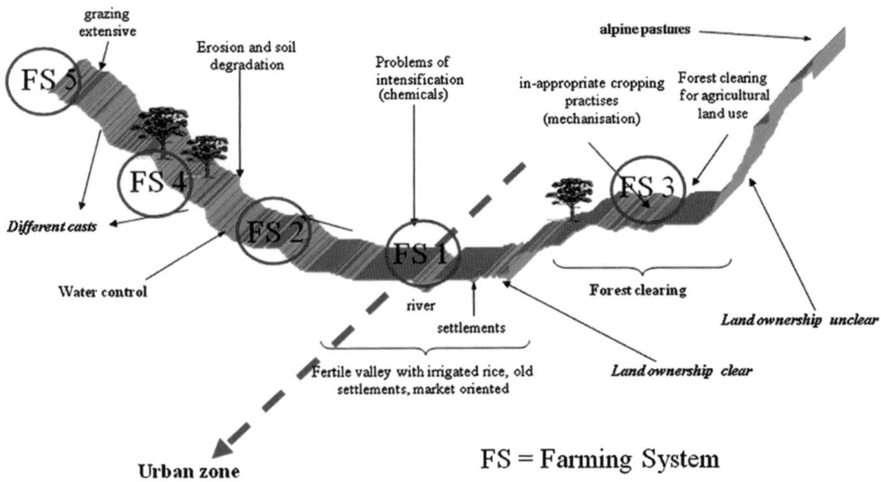

Fig. 1 Typical structure of a mountain zone valley in Himalaya region

Bhatta (2010) and Neupane (2011) have shown past developments and strategies for future actions at farm and institutional levels in mountain zones. A description of spatial zoning and farming systems in the Middle Mountains of Nepal is given by Bhatta (2010), p. 135 ff. In Northern Pakistan Muhammed (2002) designed optimal farming systems and Rabbi (2009) dealt with the developments in forest areas. In Bhutan research on farming systems development under poor conditions and restricted liberal systems is ongoing. The knowledge available from these studies together with the international literature may allow giving some rough estimates on the farming systems developments in the Himalaya region.

The main farming systems to be found in the mountainous area of Himalaya region can be summarized as follows (Fig. 1):

Farming System 1: Smallholdings with intensive rice farming systems are to be found in the fertile valley bottom at lower altitudes. Irrigation plays an important role. Land is extremely scarce; water may need investment for transportation and controlling. Livestock is small and mostly for home consumption. Market distances are usually not so big and infrastructure and transportation to urban centers are well established. Families are still considering subsistence as important but have already good relations to market sales and resource markets. Market supply may reach 30–70 % of farm production. Education is well developed; health care is at acceptable level. Administration is well developed, ownership rights are regulated and title deeds for land exist. Social life is in the frame of village communities and follows the cultural rules.

Farming System 2: Smallholdings are located at the slopes of the valley. Terraces are well developed, paddy rice is dominant, but also rainfed for other crops is practiced. Vegetable play a certain role, some of them are brought to markets. The mountain slopes are supplied by water mainly through rainfall. Much effort is needed for controlling water use and downward water flow within the group of farmers living on the same slope. On steep slopes water management coordination between farmers is an important issue to avoid erosion and land sliding. Transportation from the land to the main road may be difficult, but from main road to markets infrastructure is well developed. Subsistence farming dominates which includes crop production as well as some livestock around the house. Market relations are developed but for special products only, especially vegetables.

Farming System 3: While farming systems 1 and 2 are at an altitude between 800 and 1,800 m, farming system 3 is at a much higher level, e.g. above 2,000. In high altitude valleys we find a similar structure of farming systems at the bottom of a valley and others on the slopes. The difference to Farming systems 1 and 2 is the higher altitude with different climatic and often soil conditions, less potential for investment into natural resources and use and very long distances to markets and services. The level of remoteness is very high. Smallholdings with intensive rice farming systems in the valley bottom at high altitudes, where irrigation may be practiced but also rainfed farming with annual crops and sometimes even with fruit trees can be found. Land is scarce, but much less than in farming systems 1 and 2. Irrigation is mainly based and hand work and traditional water lifting and transporting devices. Market distances are very big and road infrastructure is limited, often carrying of products by animals or even people. Subsistence farming dominates which includes crop production as well as some livestock around the house. Market relations are developed, but may reach only 10–20 sales of the farm production.

Farming System 4: Similar to farming system 2, but much less intensive and less investment into natural resources and even more remote and less infrastructural connections to urban centers as farming system 3. Smallholdings focus on rainfed farming, but water is also used when it is easily accessible. At the sloppy farm land a variety of agricultural crops and fruits are grown. In most cases, some animals are kept, often cattle for milk or draft power. Subsistence farming is the main orientation. In some countries, special ethnic groups or casts are located in such zones.

Farming Systems 5: Livestock grazing systems in higher altitudes here crop production is reduced. When it comes to the alpine region where only simple grassland is available, the livestock is based on grazing and extensive meat production. The type of systems varies very much between different countries, especially in Tibet, Bhutan and Northern Pakistan. Sometimes milk plays a certain role for home consumption and if there is a market also for sales, in other area meat and skin/wool and other products are of relevance. Land use can reach from small to very large areas. Most of these systems can be classified as subsistence systems.

Farming System 6: Commercialized crop and livestock enterprises with relatively high capital input, external labor force, mechanization and profit oriented; usually located in valleys and close to markets or urban areas. In the mountains of Himalaya region System No. 6 is not a dominant farming system.

Table 1 Distribution of farming systems in the Himalaya region in % of resources

Farming system	Himalaya region		Nepal	
	Land	Families	Land	Families
FS 1 medium altitude, valley	28	36	25	30
FS 2 medium altitude, slopes	16	25	31	38
FS 3 higher altitude, valley	26	24	20	18
FS 4 higher altitude, slopes	10	9	16	10
FS 5 alpine region	18	5	5	2
FS 6 close to markets	2	1	3	2

The distribution of these systems in the mountains of Himalaya region has not been quantified by a large data base. Using all information available the following estimates may give a rough indication of the relevance of the individual farming systems (Table 1):

To understand the relevance of the altitudes from a settlement point of view, Nepal can be used as an example for distributions of land and people (CBS, 2002, cited from Karki 2004, p. 3). About 51 % of the households of Nepal are located in mountains and belong to the farming systems 1–5.

- High mountain: 319,887 households, population density 33 persons/settled sq.km
- Low Mountain: 1,982,753 households, population density 167 persons/settled sq.km

3 A Case from Nepal

Mountain farming in Himalaya region can be demonstrated in a case from Nepal, which is being considered as typical for this region. It stretches over a hillside down to a valley and represents a large part of the dominant farming systems in Himalaya region as shown in Fig. 1.

3.1 A Typical Mountainous Watershed

The selected study area constitutes a small mountainous watershed in mid hills of Nepal (Fig. 2). The greater part of the watershed is a mountainous under hill forest and upland cultivation. The area has a sub-tropical climate with a mean annual rainfall of 1,404 mm. The elevations of the highest and lowest point are 1,960 and 217 m above mean sea level respectively. The watershed can be divided into fertile, relatively flat valleys along the rivers (lowland) and surrounding uplands (middle land) with medium to steep slopes (highland). Agricultural land in the

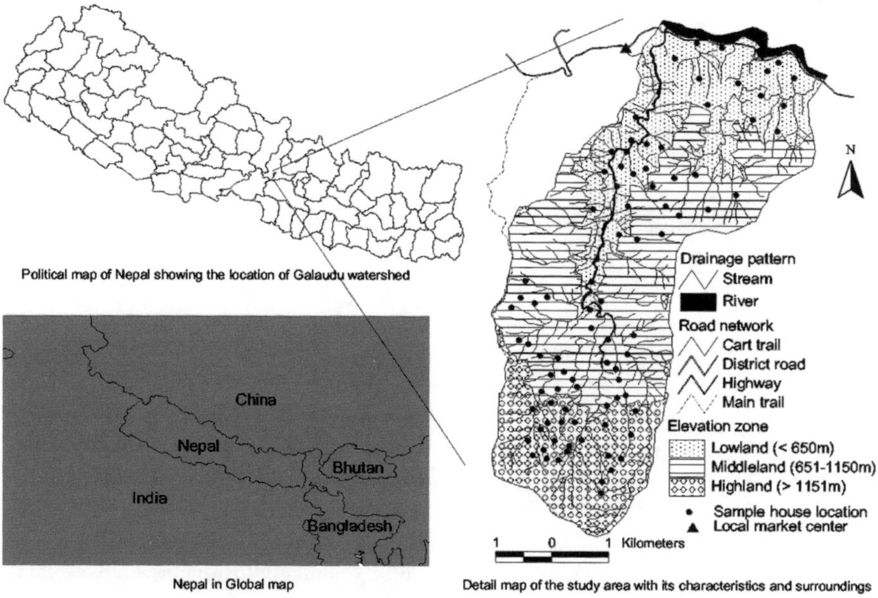

Fig. 2 Location of the study area

valleys is under intensive management with multiple cropping systems and is mostly irrigated. Paddy, potato, wheat and vegetables are major crops cultivated in the valley. Rain-fed agriculture, with or without outward facing terraces, is practiced on rest of the agricultural land, many of which is not suitable for crop production without strong soil and water conservation measures because of their high erodibility and low productivity (ICIMOD 1994).

The factors determining the different potential of development in the space and in different family allow dividing the watershed into three strata:

Highland: remote area with little infrastructure, poor access to markets, low education and monetary orientation, low production potential of land and subsistence orientation. Farming Systems can be grouped to FS 4 in Fig. 1.

Middle land: better infrastructures than in highland zone but with fewer infrastructures than lowland, reasonable access to road market, good level of education, production potential and monetary orientation than highland but less than lowland. Farming Systems can be grouped to FS 2 in Fig. 1.

Low land: relatively flat land, better access to road and markets, higher education and monetary orientation and high production potential. Farming Systems can be grouped to FS 1 in Fig. 1.

3.2 Data Base and Methodology

Socioeconomic data were obtained from family survey of spatial random sampling of 90 families using structured questionnaires. During the family survey, the geographical position of each sample household was recorded using GPS. The sample is split in three groups and farming systems respectively. Spatial data were gathered from satellite images, Global Positioning System (GPS) survey, analogue maps and digital GIS data. The spatial database consisted of land use maps derived from remote sensing. Land use maps of 1976, 1990 and 2000 were obtained by performing supervised digital image processing for satellite images of the respective years of the study area (for more details see (Bahadur 2005). This set of land use maps was used to examine land use change.

At farm family level the classical farm management analyses was carried out and extended to the level of the family. The family as the decision-making level for farm activities as well as household and all external activities has been included in the family living standard concept (Doppler et al. 2006, p. 61 ff). Family level socioeconomic and the spatial data are integrated to each other and the relation between socioeconomic and biophysical condition have been established. Socioeconomic parameter such as farm and family income has been estimated from the biophysical conditions of the area. Geographical position of each sample household was used to link the socioeconomic and spatial data and the recorded data was exported into Arc View GIS. By this process a map was obtained in which the location of all the sample households in the watershed can be seen. Subsequently, a common key field (household number) was introduced to both the point attribute table in GIS and the survey databank. With this key field, relational databases were made. In this database, the complete set of records from the sampled families was linked with their respective location. Before the constructions of thematic socioeconomic layers, spatial autocorrelation of socioeconomic variables were tested using the Moran's I and Gerry's C test and found them highly spatially auto correlated. Since the survey data was available at point level for the sampled household only, the regionalization and spatial representation required the creation of surfaces from the sample points. This was performed using Inverse weighted distance interpolation. Based on the randomly selected family's location in the area, the spatial distribution of farm and family income were prepared.

The approach adopted here to analyze the spatial relationships between over-time changes in land uses to some major ecological and economic factors, and institutional policies that are expected to have influence on changes in the watershed land use is to begin by examining the degree to which patterns of agricultural conversion can be attributed to a set of factors that have been identified as significant at broader scales in Nepal and elsewhere, namely topography, prior land use patterns, socioeconomic condition and institution governing access to land (Gautam et al. 2004; Bahadur 2005; Bahadur and Doppler 2004). The land use polygon themes for 1976 and 2000 generated from the land use assessment were overlaid in Arc View GIS Version 3.2 and location and extent of land use change were mapped and area of changes computed. The polygon theme of changes generated by overlaying the two land use themes (1976 and 2000) was then overlaid with the following GIS layers one at a time to see the spatial relationships between land use change and the respective factors including: (i) 2,500 m interval road buffers, (ii) local economy, (iii) forest governance arrangements, and (iv) socioeconomic condition.

3.3 Spatial Modeling

After the integration of socioeconomic and spatial data and creation of socioeconomic thematic layer, continuous thematic layer for both socioeconomic and spatial data are available. GIS based multiple regression model (Eq. 1) was established to estimate the spatial distribution of farm income. Based on the assumption of identical land quality in different places is expected to provide the same production and income generating potential for farms. Land quality indexes of the agricultural land uses were estimated across the watershed based on the slope, whether irrigated or not which are not only relevant for direct comparison of available land resources but also serves to relate the socioeconomic conditions, assessed by the micro survey to the physical conditions of the farm land by means of functional relations. Cost distances from the different parts of the watershed to nearest market center were measured using GIS based cost weighted distance model. The distance grid cells to travel from different location of the watershed to nearest market center were prepared. At last, the entire grid cells were combined thus both socioeconomic and biophysical condition of each and every grid cell was available together. By exporting the grid cell information to a spread sheet and then to the SPSS software package, correlations between variables were observed. Cost distance to nearest local market center and land quality parameters were found significantly highly correlated with farm incomes. Finally, multiple regression analysis was carried out by taking farm income as dependent variable and cost distance to nearest market center and land quality parameters as independent variables. Estimated income and impact maps for different scenarios were constructed by bringing back the regression result into the GIS.

$$y = 167.887x_1 - 44.385x_2 + 6946.486 \tag{1}$$

y = farm income/ha (NRS)
x_1 = land quality index
x_2 = cost distance (travelling time from and to market center in minutes)
n = 24047 grid cells $R^2 = 0.729$
F test = 13508.786 sig. F = 0.000
T-Stat for constant = 186.642 p value = 0.000
T-Stat for coefficient of land quality index = 14.866 p value = 0.000
T-Stat for coefficient of cost distance = −145.196 p value = 0.000

The features of this GIS based multiple linear regression model indicated a good explanatory value of the relationship with a measure of determination (R^2) of 0.729 and sufficiently high levels of significance for the whole function (F-test) as well as its components (t-tests), which exceeded a probability level of 99 % in all cases. The model aims at regionalizing the current income situation and uses statistical dependencies for the simulation of the effects of future strategies. In general, the transfer of the estimation to all the sample location showed reliable results. In highland areas, income declines with the distance from road and market center, higher elevation with upland areas. These low-income zones are reflecting the combined effect of remoteness and the less favorable land conditions. The high-income areas are located relatively near to the main road, local market center, at lower elevation areas of valley bottom where as low incomes areas are located more on the hilltops at higher elevation, mainly steep slopes, far from the road and market center. This difference again reflects the resource (especially land) quality of the areas and their connection to market center through road networks.

3.3.1 Application of Model for Future Strategy Testing

The GIS based multiple regression model used for income estimations was used to estimate potential future income generation in different scenarios of farm management. For this purpose, the land quality index of the grid cells and cost distance from the each of the grid cell to the nearest market center were modified from the current situation. The farm management model is based on the changes of the land quality index according to the requirements of the defined scenarios. The value of the land quality index depends on the terrain slope, state of agricultural land whether irrigated or not, of the grid cells (Bahadur 2005). Slope cannot be modified, what can be modified is the weight associated with the state of land whether irrigated or not, soil and nutrient losses according to the slope (Bahadur 2005) and practiced or assumed management of the land (Lentes 2003). Since the index represents the differences in the quality of the land, different land qualities can be simulated with the model by changing the weighting factors of the individual classes as required for the setting up of scenarios. Modifications of the weights of individual grid cell of different themes allow the simulation of future land quality

Table 2 Farm and family income by sub study zone, 2003

Income (NRS)	Lowland	Middle land	Highland	Overall
Farm*	35,396	29,498	8,228	24,375
	(27,523)	(27,940)	(4,874)	(25,428)
Off farm**	8,283	5,614	4,781	6,226
Family*	43,679	35,113	13,011	30,601
	(34,607)	(27,750)	(6,748)	(28,719)
Per person*	6,835	4,730	2,219	4,595
Per labor unit*	16,844	12,211	6,061	11,705

NRS = Nepali Rupees (1US$ = 80 NRS) figure in parenthesis are standard deviation
* and ** = Significant at 99 and 90 % respectively
Source Bahadur (2005), p. 167

index of the respective grid cell. The final land quality index for each grid cell is calculated by multiplication of weight given to each of individual grid theme. Higher land quality index signifies better land quality. In the multiple linear regression models, this was used alone and together with the cost distance to explain the farm income.

In future farm management model construction, consideration was given for example what would happen if the given weights are modified to match the desired scenario. If the final weight of a grid cell increases for example from 1 to 1.1, the higher future land quality index for the given cell can be expected. Likewise, if the final weight of a grid cell decreases for example from 1 to 0.9 than smaller value of future land quality index for the given cell can be expected. Cost distance values after the improved road scenarios were used for simulating the effects of improved infrastructure on the income/ha of the sample household. For this purpose, the functional relationship found in the current situation was applied with the new cost distance values. The new income/ha for each and every grid cell was estimated and compared with the current situation and differences were taken as the impact of improved road.

3.4 *Analyses of Current Situation: Farm Family Income and Spatial Distribution*

The results of farm, off farm and family income show in crop year 2002/2003, farm income was 35,396; 29,498 and 8,229 NRS in lowland, middle and highland areas respectively (Table 2). Average off-farm income was 8,283; 5,614 and 4,781 NRS in lowland, middle land and highland respectively. The average annual family income was 43,679; 35,113 and 13,011 NRS in lowland, middle land and highland respectively. According to Kruskal–Wallis test, there were statistically significant differences between sub-study zones in farm, off farm and family incomes (probability of 99 %). A pair wise test of significant differences between two sub-

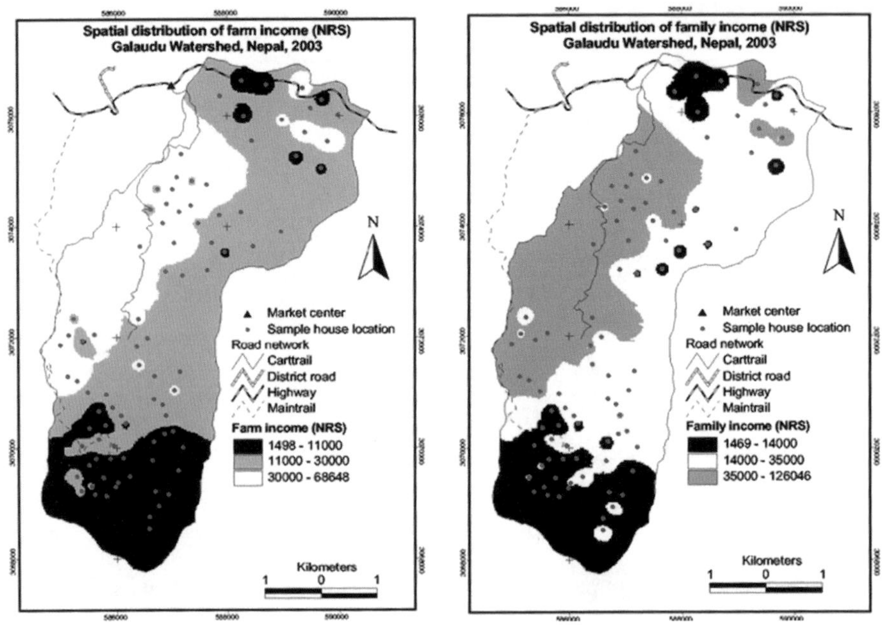

Fig. 3 Spatial distribution of farm and family income. *Source* Bahadur (2005), p. 130

study zones in family income (Mann–Whitney Test) showed that lowland and highland and middle land and highland were highly significantly different but lowland and middle land did not differ so much in the level of family income.

The spatial distribution of farm and family incomes are presented in Fig. 3. Income differentiation in the space shows a higher farm and family income in the most favorable zones and near to road and market center. The spatial differences in farm income between these sub-zones are mainly due to the proportionally smaller size of irrigated land per family in highland areas and their productivity as compared to lowland. Besides, vegetable farming, which is intensively practiced in lowland areas, could be another reason for having high farm incomes at lowland areas. This was also confirmed from the higher farm income per ha of cultivated land and per person from the lowland areas as compared to highland zone (Table 2). Farm income was more or less similar in lowland and middle land. Families in these areas were able to earn similar levels of farm income because of well-terraced land, more intensive use of external inputs and crop diversification allows crop rotation and reduces risk of crop failure. Farm income accounts for greater share of the family income in all three sub study zones. Family income per person and per labor unit were significantly higher at low land as compared to highland which could be due to the accumulated effect of land quality, education level of farmers, distance from market and road.

3.5 Analyses of Current Situation: Land Use Changes and Reduction of Forest Areas

To characterize the volume and quality of land resources available and even more the decision-making of people who use the land, the development of land uses over time as well as differences according to the location in the area and the conditions of the locations in the space is to be understood. During ten year periods from 1990 to 2000 forestland declined by 10.6 % while upland agriculture increased by 9.6 % and lowland agriculture by 1 % (Fig. 4). The annual rate of forest loss in the study area was about 1.1 %. Land uses change may be attributed to the spatial location of land.

The change of land use along altitudinal gradients in determining the type of forest vegetation occurs at different physiographic regions across Nepal has been widely documented (e.g. Jackson 1994). Little is known, however, about the association of altitudinal gradients with changes over time in forest cover. The conversion of forestland to agricultural activities was not similar throughout the watershed (Fig. 4). In the highlands zone more forestland was converted to agricultural land as compared to lower elevation (low land) area. Overlaying a polygon theme of sub study zones with polygon theme of land use changes during 1990–2000 showed that higher elevation forests were more dynamic compared to lower elevations. The rate of forestland conversion to agricultural activities was at least two-times higher compared to locations at lower elevation (lowland) areas (Table 3). Around 36 % of the forest area in 1990 within the higher elevation zone (highland) was converted into agricultural land where as about 18 % forest area from middle altitude and only 6 % forest area from lower elevation (lowland) was converted into the agricultural land in the same period.

The effect of accessibility on the changes in forest area reflected increasing conversion of forest area into agricultural activities with the increase distance from roads. 70 % forest area in 1976 within 2,500 m distance from roads remained unchanged until 2,000, whereas only 44 % forest areas located 5,000 m far from roads were unchanged during the same period (Table 4). Proportionately lower amount of forest loss in areas of better accessibility, however, is generally an unexpected trend. Higher concentration of forest management activities in locations closer to the roads might be a reason. Likewise effective monitoring of the community forests by local user groups could be another reason for improved forest condition in relatively accessible areas (Gautam et al. 2004).

Forestry requirements and forest management objectives of more market oriented farming systems are different from those of more subsistence based systems. In the first case forest management objectives are mainly for watershed protection. An overlay of polygon theme prepared by creating a 2,500-m buffer from the local market center with the land use polygon theme of changes in forest area during 1976–2000 revealed that more forest area was converted from forestland to agricultural activities in the more subsistence systems than in more market oriented areas. (Table 5).

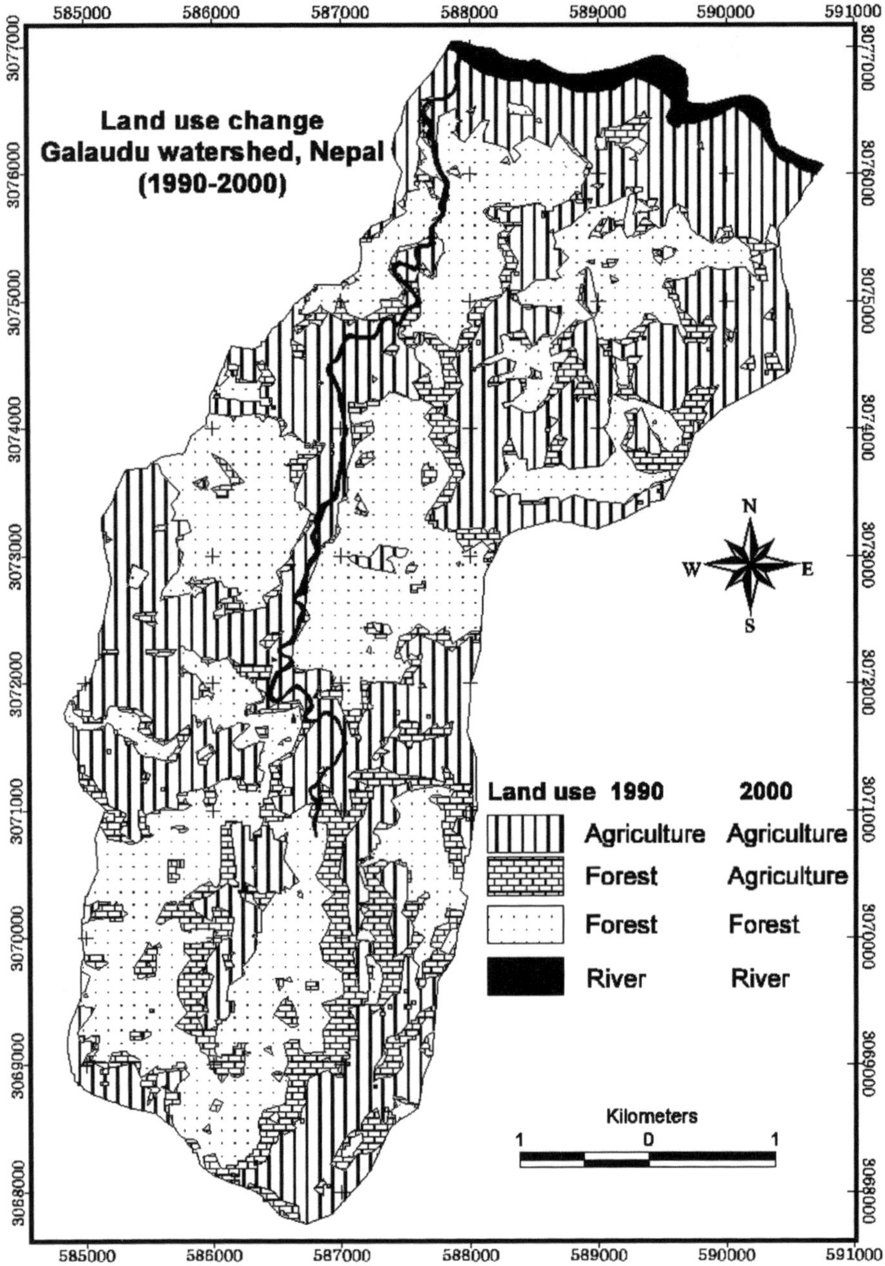

Fig. 4 Land use change in the period of 1990–2000

Table 3 Change in land uses by sub-study zone in between 1990 and 2000

Land cover	Lowland			Middle land			Highland		
	1990	2000	Change	1990	2000	Change	1990	2000	Change
Forest (ha)	395.8	371.8	−24	688.9	566.9	−122.0	391.2	251.0	−140.2
Upland agriculture (ha)	76.8	95.8	+19	239.4	353.6	+114.2	98.5	221.8	+123.3
Lowland agriculture (ha)	360.6	365.4	+4.8	365.9	373.8	+7.9	83.3	100.2	+16.9

Table 4 Change in land use by accessibility in between 1976 and 2000

Road buffer (m.)	Forest (ha)		Lowland agriculture (ha)		Upland agriculture (ha)	
	1976–2000	1990–2000	1976–2000	1990–2000	1976–2000	1990–2000
<2,500	−194.0	−38.1	+193.2	−3.5	−5.6	+41.0
2,500–5,000	−124.0	−30.3	+95.5	+9.6	+26.6	+19.8
5,000–7,500	−253.0	−146.0	+124.0	+11.6	+129.3	+134.6
>7,500	−64.4	−73.3	−15.2	+11.6	+79.4	+61.6

Table 5 Change in land use in relation to subsistence and market orientation of farming systems between 1976 and 2000

Farming System	Forest (ha)		Lowland agriculture (ha)		Upland agriculture (ha)	
	1976–2000	1990–2000	1976–2000	1990–2000	1976–2000	1990–2000
More market	−89.6	−98.5	−15.6	11.6	105.9	87.0
Medium market	−246.0	−126.0	137.7	15.2	106.8	110.6
Little market	−225.0	−63.0	185.5	13.7	35.9	48.0
Subsistence	−73.7	−0.3	89.8	−11.1	−18.9	11.4

According to Gautam et al. (2004) areas with formalized community forests had significantly higher shrub lands-to-forest conversion during 1978–1992 compared to areas without formal community forests. The fact that proportionately less amount of forest lost and degraded those were managed with the involvement of local forest user groups supports the argument that legal transfer of resource ownership is an important precondition for successful collective outcomes at the local level. One of the distinct differences between community forests and government forests in this watershed is the involvement of local communities in forest conservation in the former case. From this point of view, the findings of this study indicate the joint investment by local forest users and local agencies may improve the prospects for successful forest conservation at local level (Gautam et al. 2004). Conversations with local forestry staff and local people revealed that the forested areas under the government control are virtually in “open access” condition as the district forestry staffs are mostly engaged in community forestry activities after the implementation of community forestry program. So the relatively high loss of forest area under state control can be explained by their condition of open access.

3.6 Future Strategies Assessment

The future of mountain agriculture for zones in the Himalaya region similar to this study area can be simulated by the assessment of several future strategies relevant for development. According to the most determining factors for development in this region land and water resources management as well as road infrastructure strategies have been selected.

3.6.1 Improved Land Management

Results of the improved land management on income are presented in Fig. 4. The income generating potential is increased through improvement of land management. After the implementation of assumed land management scenarios, simulated income will not be distributed evenly throughout the study area. Simulated income will still be the highest in the current highest income areas and in the area where the impact of changed management is estimated to be highest. As in the current situation, the high-income areas remain same as they were before. Nevertheless, the situations in the low-income zones, especially in highland zone were changed. While the current income in much of the highland zone is below 20,000 NRS, the improved situation predicts the income per hectare to reach and exceed 25,000 NRS (Fig. 5). The percentage increase of income with soil conservation measures compared to current situation can be seen in Fig. 5. The low-income areas in highland zone benefit most from the changed management.

3.6.2 Improved Road Network

Results of the improved road network on spatial distributions of the income in the watershed are presented in Fig. 6. Improved cost distances were used for simulating the effects of improved infrastructure on the income/ha. On spatial level, improved situation show the great differences over the current situation. Significant impact on income of farmers was observed before and after the improvement and development of road networks in the remote areas. The impact of improved road was the highest in the remote areas where there were currently no infrastructural networks. Still, the level of income is low in this area, when it is compared to the high-income areas. In the zone with the current lowest income (20,000 NRS/ha) the scenario predicts the highest increase. For the currently high-income areas, the increase of income is predicted to be less than 16 %. Better response is found in the rest of area of middle altitude and most area of the highland zone, where 20–96 % increase of income is predicted increasing with the distance from the marketplace. Similar results from Middle Mountain sof Nepal are shown by Bhatta (2010) p. 192 ff).

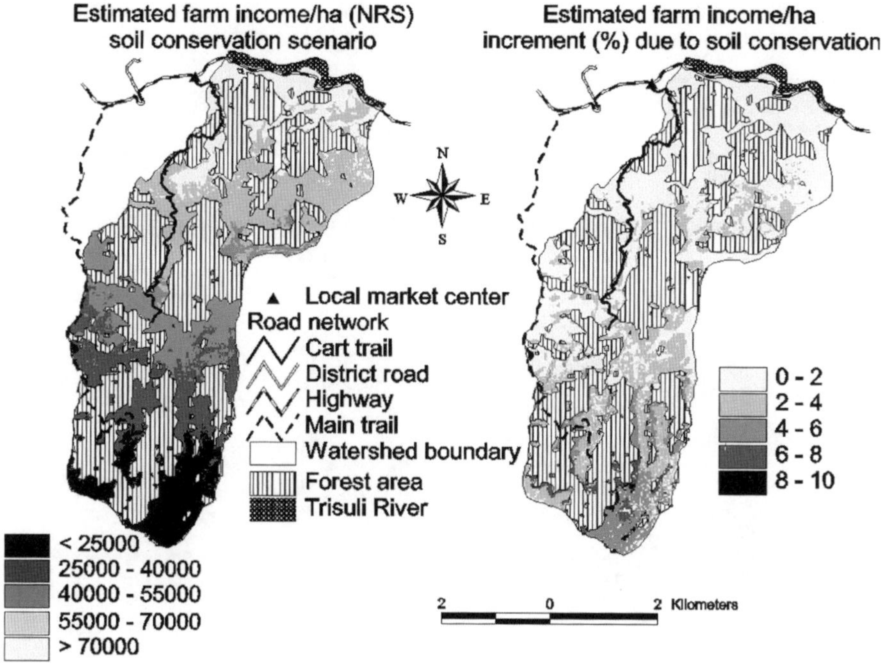

Fig. 5 Assessment of the future strategies of soil conservation on farm income: Simulated farm income with soil conservation (*left*) and impact of soil conservation on income (*right*). *Source* Bahadur (2005), p. 191

3.6.3 Combining Strategies of Land, Water and Road Improvement

Significant impact on incomes was observed by improving water resources management especially in the highland zone. The impact of improved road was the highest in the remote areas. Since these areas are also identified as those with the highest potential for income boost through the introduction of improved land and water resources management activities, the combination of more than one measure yields the best response (Fig. 7). Still, the level of income is low in this area, when it is compared to the high-income areas. In the zone with the current lowest income (20,000 NRS/ha) the scenario predicts the highest increase. For the currently high-income areas, the increase of income is predicted to be less than 16 %. Better response is found in the rest of area of medium market relations (middle altitude) and subsistence farming (most area of the highland zone), where 31–104 % increase of income is predicted in the case of combined land and road improvement and 30–143 % in the case of combined water and road improvement. Nevertheless, the impact of water resources management on the sustainability of the farming system is expected to be substantial in these zones, since the cultivation of slopes induces heavy soil loss in this area too.

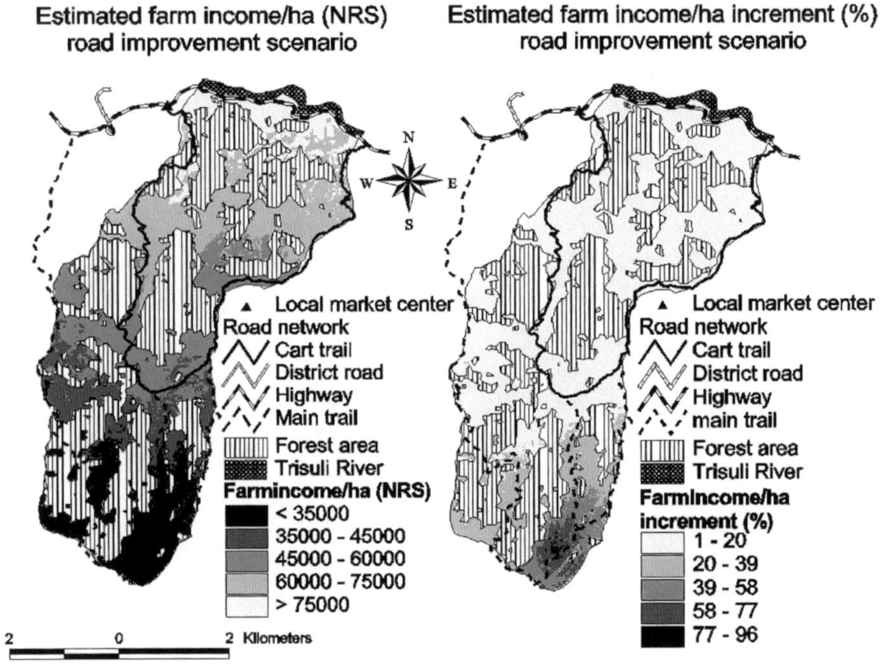

Fig. 6 Assessment of the road improvement on farm income: simulated farm income with road improvement (*left*) impact of road improvement on income (*right*) *Source*: Bahadur (2005), p. 197

The overall results show that qualities of agricultural land and their management, education level of settlers, off farm employment opportunities, consumption of family labor, food security, and forestry requirements of residents living in highland areas are different from those living in lowland areas of the watershed. Residents living in lowland area have alternative sources of energy for cooking and heating, most of the households have better agricultural land, better off farm opportunities being near to road and market center compared to those in highlands.

There is a spatial relationship between resource use, degradation and socioeconomic status of people in the different spatial conditions of the watershed. Spatial distribution of farm income, family income, crop production and percentage of food bought show as the elevation increases, distance from road and market center increases, farm family incomes and crop production decrease. The percentage of food bought, as an indicator of food security supply for the family, increases. This shows the difference in the spatial distribution of socioeconomic attributes is due to the difference of accessibilities of road and market centers and the management of the available agricultural resources. Priority should be given to the policies relevant for creating better access to road and market centers from each corner of the watershed to improve the living standards of farmers and reduce spatial differences between the low and high land watershed zones.

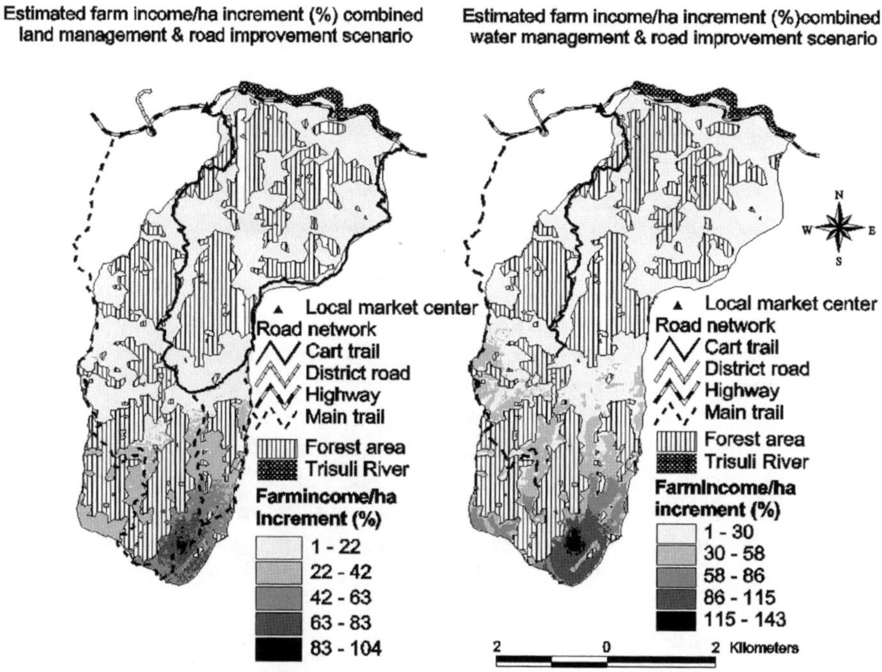

Fig. 7 Assessment of the combined strategies: Impact of combined land and road improvement on income (*left*) Impact of combined road and water improvement on income (*right*)

The quantitative evidences of land use dynamics, which were delivered by a time series of satellite images coupled by GIS analyses, collaborate the findings of some earlier studies (Schreier et al. 1994; Virgo and Subba 1994; Jackson 1994) that the deforestation trend in some areas of the Middle Hills of Nepal have continued even though governmental and non-governmental organizations have been carrying out different community forestation programs. One important change was the increase in agricultural area at the cost of forestland possibly due to the expansion of settlements. Even though it is only in the lower elevation area of the watershed, the positive changes in forest cover, probably due to the community forestry management programmes, provides some evidences of ecological sustainability of the resource at the lowland area of the watershed, although the reversal of the decreasing trend in scrublands during 1990–2000 period has raised some questions regarding the possible continuation of the observed trends in future. Some other important concerns related to community-based forest management in Nepal are: (i) whether and how the positive changes in forest cover has benefited the local users, and (ii) how sustainable are the existing community-based forestry institutions in the long run.

4 Lessons for the Himalaya Region

Mountain farming in Himalaya region has much in common with many other mountain regions in the world. One main common feature is the settlements along valleys and respective slopes and mountains with the consequence of special land and water use conditions, the connection to forests with the pressure of cutting forest for agricultural land use and changing socioeconomic and cultural conditions the more remote the area is.

A common strategy for improving living standard in mountain regions is the transition from subsistence to market oriented farming systems. This can be defined at administrative level and implemented through different types of projects or extension concepts and activities as discussed in many regions in the world. For all different levels of subsistence and market relations it is equally important to bring them closer to markets. In pure subsistence families this is a very big task and will also mean to change the way of life. For that reason, one has to expect a time consuming process. Very often, the initiative will have to come from outside. Such outside strategies may be related to governments in building road infrastructure, in providing information and knowledge about the potential of agriculture. This may also be supported by inside effects when people realize the potential for a better life through information systems and media of different types.

There are important *special features of Himalaya mountain regions*. The future potential has been investigated in several research projects at the universities of Hohenheim and Giessen from 2000–2011 (see Chap. 2) and shown that there is a great potential for development, but strongly related to the danger of man-made problems. This requires special consideration and guidance for development. The four main sectors of future development will be summarized as follows.

4.1 Development Concepts Adjusted to Socio-Cultural Background

Traditional cultures and religious impacts have created a certain way of life and influence the decision-making in farming families as well as in small societies in settlements. Social stratifications like the caste system have an additional impact on possible social groupings in the area and affecting ownership and user rights, linkage to resources and business and influence in heritage and marriage rules in farming families. For these reasons any institutions working in these areas will have to consider this in their concepts and approaches (Doppler 2012, Sect. 4.1).

- There is a strong need to overcome the partial approach of farm efficiency thinking to the family decision-making. Especially pronounced in subsistence farming are multiple objectives where not only economic objectives but also socioeconomic and cultural interests are influencing farm families' decisions.

Living standard analyses are more adequate (for details see Doppler et al. 2006 and Doppler 2012). Families may have to decide on farm, household and off-farm allocation of resources (horizontal dimension) and will have to consider the level of dependencies from resource owners (land, capital, labor etc.)

- With increasing relation to market there is an increasing relevance to include the interdependencies between micro, meso and macro level. The focus is on the vertical dimension between families and institutions and organization in the region and in the country who decide on resource availability, legal ownership rights, marketing and capital flows. This is especially relevant in the downstream water control infrastructure and management.
- The development over time (dynamic dimension) in mountain areas is especially relevant in the context of long-term (sustainable) regimes of natural resources with special relevance to lands and water.

4.2 Road Infrastructure to Connect to Markets and Service Centers

Road infrastructure is one of the driving forces for development in the Himalaya mountain region. It connects people to areas and centers from where they may get products, resources, external knowledge and legal support and in which they may provide their products, local knowledge and labor force for the supply in more developed zones. Furthermore, cultural relations and exchange may bring people together and influence their social and political relations in long-terms. This is especially pronounced in Northern Pakistan, Northern India and Tibet and has also an impact on the relation between culturally different, ethnic and religious groups such as mirrored by the caste system.

4.2.1 Connecting Remote Areas to Urban Centers

Since families (farms), villages, and markets are located in a region determined by its distances in location, there is an urgent need for including the spatial dimension in relation to settlements, infrastructure and institutions/organizations. As has been shown in Chap. 3 the impact of improving roads infrastructure will bring a big contribution to rural development in mountain zones. This has also been confirmed by other research projects of Lentes (2003), Rudiarto (2009) and Bhatta (2010). While in large plain areas urban centers may grow at any place (mostly close to rivers or big roads) and farming and business will develop around such centers, in mountain regions urban centers are usually developing in the highest hills, but downwards rivers in adjacent plain areas. This indicates the most relevant ways of development in the mountains: connecting farming families in remote zones to the urban centers in plain areas or at the beginning of large valleys.

4.2.2 Interrelationship Between Valleys in Many Stages Climbing up the Mountain

In the Himalaya region there are special spatial conditions for improving road infrastructure. Due to the high altitudes, one may find several valleys following each other on the way from 800 to 5,000 m altitudes. This means each valley at a higher level starts at the end of an upper valley and floors into a valley at a lower stage. This stage structure of valleys also means that road infrastructure for upper valleys depend on the road infrastructure at lower level. It also means it does not help much to develop the road system at higher level, when the intention is to reach the urban centers down at low land regions and at a lower level of a valley the road infrastructure is not adequate. As a consequence, there are very long distances from the settlements of farmers in remote areas to urban centers in flat areas. Sometimes there may be even six valleys following each other downward to urban centers. This is why there is an extremely poor road infrastructure at higher altitude valleys and high investment cost occurs to improve road infrastructure in areas where population density is low and the efficiency of road investment cost per person is even lower. Large areas in remote zones are currently only connected to outside markets by people or robust animals carrying household goods and farm inputs to the families and farm products to markets.

4.2.3 Road Infrastructure and Mobility

As many simulations show (e.g. Lentes 2003; Rudiarto 2009), improving road infrastructure has two effects of developments: it connects farming to markets, but also offers transportation of people to other places. In a future assessment of the impact of road improvements in high mountain zones the mobility of people will be of central importance with respect to daily mobility due to better jobs for off-farm income as well as for moving to other places with the families for settling down. As far as people are still in a traditional culture the migration to other places may not be so interesting for them. The more information is available and traditional cultures are neglected, young people may move to other locations if job opportunities there are expected to be good. Different to other mountain zones in the world, in Himalaya region the cultural ties and the tradition in many remote zones are still strong enough that massive movements from high mountain areas to cities will not yet take place. In future such developments may get higher importance.

4.3 Down-Stream Water Control Infrastructure and Management

The amount of rainfall in the Himalaya region is very high. A lot of water runs down the mountains. Many farming systems depend on this water. At the same time the large amounts of water running down are also subject to damage of terraces as well as of larger parts of hills and slopes. Without controlling water flows in fields and rivers through very long ways from highest mountains to the plain areas settled and cultivated land cannot be protected. In the Himalaya region there is a big tradition in controlling such water. In areas where quick settlement takes place after cutting of forests the time often does not suffice to establish a down-stream water control infrastructure or management. In such cases soil erosion, breakdown of terraces or land sliding is the big threat.

In a situation where one valley follows another valley up the mountains, the improvement of road infrastructure of the highest valley depends also on the road infrastructure of the valleys downward to connect to flat areas development. It is advisable to develop road infrastructure from the lowest part to the top. With the control of water it is just the opposite direction. Control of water has to start at the highest level where water is appearing in larger amounts. The control and management of water has to be organized downward, farmer by farmer, terrace by terrace, and river by river. Rudiarto (2009) and Doppler et al. (2009) has shown the consequences of a disastrous development especially in newly settled areas. The fact, that in Himalaya region these problems are not as strong as in Southeast Asian mountains is due to the long traditional experiences in Himalaya region, but will occur when new settlements take place in a short time on long slopes of mountains.

4.4 Institutional Infrastructure

The higher the altitude in the mountain zone and the more remote the area is the more often lack of institutional capacity can be found. This is true for administrative, social and economic infrastructure.

4.4.1 Special Relevance for Institutional Development

The development of a proper institutional infrastructure depends much on the staff of institutions available in remote areas. Mostly they will come from developed regions with good knowledge and the supply of such institutions and organizations also needs direct relation to developed areas. The main challenges of institutional development are as follows:

- Lack of monetary markets and regulated market structure including proper financial institutions and regulated credit markets.
- Lack of an administration which takes care of legal needs in resource use and ownership of land (title deeds on land), water rights and regulations of distributions and capital and credit issues.

- Lack of a proper education system, even primary school may be available; the potential for higher education is extremely limited. This includes also education and training in professional fields.
- Lack of proper health care at a basic level and even more at a higher level. Distances to hospitals may be as far as days from the homestead.
- Information systems to transfer knowledge. An urgent need for development in remote areas is the transfer of knowledge in professional fields like, farming, business, education, health care etc. Changes through the introduction of electronic communication systems have been induced. It also has to be expected, that this will influence the objectives and values of traditional societies in remote areas in long-term.

Migration is often not only due to economic opportunities but also to better education, health and social life. In many interviews the male population indicates the importance of the economic opportunities and that they are ready to migrate, but women often very clearly speak about their preference for educating their children for a better future and for improving the health of their family. Development in the remote zone will have to include these issues as much as any other issues like road infrastructure or low-level agricultural production. Education and health is not only a strategy to keep people in their home area but also to improve their living standard in general.

4.4.2 Acceptance of Regional Unbalanced Development

Investments into linking remote areas to urban centers are extremely expensive in mountain zones and in the Himalaya region even more expensive than in other parts of the world. Required investments need to be restricted to selected problems and regions in a given time since financial resources are limited. It is for that reason that mountain farming development will take place in one zone more than in another. Often a single investment is not enough in a region (e.g. schooling infrastructure) but a set of investment is needed (road, communication systems, schooling, health care and legal sector administration) to reach the success required. At the same time, investments may have to start at valleys of lower levels, move upwards onto mountain areas (e.g. roads) and may call for priority development for valleys closer to urban centers. This means that regionally unbalanced development is justified and may be the only realistic alternative and only in long-term views there might be a certain balance of development. Sometimes different tribes, casts or ethnic groups may live in different zones of development and may experience political preference or restrictions. Under such conditions unbalanced development may create political tensions.

5 Conclusion

The final question, whether mountain farming will continue to exist and play an important role in the future is clearly to be answered with "yes", but many farming systems and regional structures will change. The growth of population will lead to increasing scarcity of land and to the danger of inadequate use of the natural resources land and water. It can even be expected that in the forest zones the pressure for transferring forests into agricultural land will grow. Man-made problems will increase. The potential for improving living standard of people is big and the potential for increasing the number of families in remote zones does exist. The infrastructure in roads, electricity and drinking water supply systems, as well as communications systems is basic for production, business, and social life and for education (school systems) and health care in rural areas. Development of infrastructure is a precondition for a sustainable increase of living standard of the current population as well as for families settling down in this zone in the future. The responsibility of the public sector is increasing strongly to ensure proper use of natural resources and increase living standards, but the preconditions for this are very hard: strict control of resource ownership and use, efficient organization and management of water control, road and social infrastructure developments and an increasing liberalization of society and social organization.

References

Bahadur KC K (2005) Combining socioeconomic and spatial methodologies in rural resources and livelihood development: a case from mountains of Nepal. In: Doppler W, Bauer S (eds) Farming and rural system economics, vol 69. Margraf Publishers, Germany

Bahadur KC K (2005) Modeling farm income using remotely sensed data and GIS in a rural mountainous watershed in Nepal. In: proceeding of GIS planet 2005 (II international conference and exhibition on geographic information system). Estoril, Portugal. ISBN 972-97367-5-8

Bahadur KC K, Doppler W (2004). Integrating micro level and remote sensing data in GIS analyses for natural resources management and socioeconomic development. In: Proceeding of international conference on rural poverty reduction through research for development and transformation. Humboldt University, Berlin, Germany. http://www.tropentag.de/2004/abstracts/full/115.pdf

Bhatta GD (2010) Socio-economic and spatial assessment of smallholder peri-urban farming in the Middle mountains of Nepal. In: Doppler W, Bauer S (eds) Farming and rural system economics, vol 117. Margraf Publishers, Germany

Bhatta BR, Chalise SR, Myint AK, Sharma PN (eds) (1999) Recent concepts, knowledge, practices, and new skills in participatory integrated watershed management: Trainers resource book. ICIMOD, PWMTA-FAO, and Department of Soil Conservation and Watershed Management, Nepal

Das R (2010) Bio-economic modeling of soil resource conservation in mid hill region of Nepal. In: Doppler W, Bauer S (eds) Farming and rural system economics, vol 110. Margraf Publishers, Germany

Doppler W (2012) Tropical farming systems economics. History and impacts of a chair at the University of Hohenheim, Germany. Margraf Publishers, Germany
Doppler W, Praneetvatakul S, Munkung N, Sattarasart A, Kitchaicharoen J, Thongthap C, Lentes P, Tai D, Grueninger M, Weber K (2006) Resources and livelihood in mountain areas of South East Asia. Farming and rural systems in a changing environment. Margraf Publishers, Germany
Doppler W, Bahadur KC K, Kitchaicharoen J (2009) Water resources development and its impact on rural livelihood in Northern Thailand—Integrating GIS into farming systems economics. Margraf Verlag, Weikersheim, p 240
Gautam AP, Shivakoti GP, Webb EL (2004) Forest cover change, physiography, local economy, and institution in a mountain watershed in Nepal. Environ Manage 33(1):48–61
Goshi J (2005) Variety attributes farmers' preferences and rice technology adoption in Nepal. In: Doppler W, Bauer S (eds) Farming and rural system economics, vol 75. Margraf Publishers, Germany
ICIMOD (1994) Application of GIS in rural development planning in Nepal. International Centre for Integrated Mountain Development (ICIMOD), Kathmandu
Jackson JK (1994) Manual of afforestation in Nepal. (in two volumes) Forest Research and Survey Center, Kathmandu, Nepal
Karki LB (2004) Impact of project intervention on rural households in Nepal: assessment os socio-economic and environmental implications. In: Doppler W, Bauer S (eds) Farming and rural system economics, vol 60. Margraf Publishers, Germany
Lentes P (2003) The Contribution of GIS and remote sensing to farming systems research on micro-and regional scale in North west Vietnam. In: Doppler W, Bauer S (eds) Farming and rural system economics, vol. 52. Margraf Publishers, Germany
Maharjan A (2010) Labor out migration and its impact on farm families in mid hills of Nepal. In: Doppler W, Bauer S (eds) Farming and rural system economics, vol 113. Margraf Publishers, Germany
Muhammed R (2002) Optimal agricultural land use systems for Northern Pakistan. determination through quadratic risk programming. In: Doppler W, Bauer S (eds) Farming and rural system economics, vol 40. Margraf Publishers, Germany
Neupane N (2011) Political-economy of water distribution in the trans-Himalayan region of Nepal. In: Doppler W, Bauer S (eds) Farming and rural system economics, vol 128. Margraf Publishers, Germany
Rabbi F (2009) The economic reliance and degradation of forests in Northern Pakistan. A socio-economic analyses. In: Doppler W, Bauer S (eds) Farming and rural system economics, vol 106. Margraf Publishers, Germany
Rudiarto I (2009) Sustainable land use concept in mountain area of Java, Indonesia. In : Doppler W, and Bahadur KC K (eds) Farming and rural systems economics and biodiversity in the tropics. Proceedings of the international DAAD alumni summer school, Sep 29–Oct 4, 2009, University of Hohenheim, Germany. Magraf Publishers, Weikersheim, pp 1–29
Schreier H, Brown S, Schmidt M, Shah P, Shrestha B, Nakarmi G, Subba K, Wymann S (1994) Gaining forest but losing ground: a GIS evaluation in a Himalayan watershed. Environ Manage 18(1):139–150
Virgo KJ, Subba KJ (1994) Land-Use Change between 1978 and 1990 in Dhankuta District, Koshi Hills, Eastern Nepal. Mt Res Dev 14:159–170
Von Thünen JH (1826) Der isolierte Staat in Beziehung auf Landwirtschaft und National-Ökonomie: Untersuchungen über den Einfluss, den die Getreidepreise, der Reichtum des Bodens und die Abgaben auf den Ackerbau ausüben

About the Authors

Kamiljon Akramov Research Fellow of IFPRI, received PhD in Economic Policy Analysis from RAND Graduate School, USA in 2006. His research focuses on aid effectiveness, institutional change, agricultural growth and food security in Central Asia, and decentralization and efficiency of service delivery in transition and developing countries. He has worked at earlier stages at RAND Corporation and at the Central Bank of Uzbekistan.

Jeffrey Alwang is professor of Agricultural and Applied Economics at Virginia Tech. His research focuses on policies to alleviate poverty in rural areas, sustainable agricultural growth for poverty reduction, and rural development. Recent international research projects have focused on poverty reduction through sustainable agricultural development in the highlands of Ecuador and Bolivia. He has received research funding from, among others, USAID, the World Bank, USTDA, FAO, UNICEF, and UNDP.

Sarah Cline is an Economist at the U.S. Department of the Interior. She holds a PhD in agricultural and resource economics from Colorado State University. Her research has focused on natural resource and agricultural economics, including a study of agricultural landscapes in Colorado's central mountain region. Sarah has previously worked at the U.S. Department of Agriculture and the International Food Policy Research Institute.

Xinshen Diao Senior Research Fellow of IFPRI, received PhD from Department of Applied Economics, Universality of Minnesota, USA in 1995. Xinshen has long time experience working on economic development and growth, intersectoral linkages, international trade, and regional integration and dynamics. Xinshen is a dynamic general equilibrium modeler and has published more than 80 articles in various economic journals, IFPRI Research Reports, IFPRI Research Monographs, and economic books.

S. Mann (ed.), *The Future of Mountain Agriculture*, Springer Geography,
DOI: 10.1007/978-3-642-33584-6, © Springer-Verlag Berlin Heidelberg 2013

Werner Doppler has been a professor for farm management in tropical regions at the University of Hohenheim between 1981 and 2008. Officially retired since then, he is still actively doing research in Nepal about development projects in the Himalaya. Werner Doppler has published 234 peer-reviewed papers and supervised 57 PhD students.

Christian Flury leads the Research Programme AgriMontana “Agriculture’s contribution to sustainable development in the mountain area” at the Federal Research Station Agroscope in Switzerland. He holds a PhD in Agricultural Economics from the Swiss Federal Institute of Technology Zurich. Besides his work at Agroscope, he is co-founder and now partner and consultant in an agricultural and regional economic consulting firm.

Jonathan Hellin works at the International Maize and Wheat Improvement Center (CIMMYT) in Mexico. He received his BA and MSc degrees from the University of Oxford, and his PhD from Oxford Brookes University. After completing a cross-disciplinary PhD on smallholder land management in Central America, his research has focused on farmers’ livelihood security and access to markets. He has authored two books and over 40 peer-reviewed journal articles and book chapters.

Feliu López-i-Gelats is a postdoc researcher at Centre for Agro-food Economy and Development in Barcelona. He holds a PhD in Environmental Sciences. Before, he worked at Food and Animal Science Department and at Institute of Environmental Sciences and Technology in Autonomous University of Barcelona. He has also been Marie Curie research fellow in Keele University and in Aalborg University.

Stefan Mann heads the Socioeconomics research group at the Federal Research Station Agroscope in Switzerland. He holds a PhD in Economics and a second PhD in agriculture. Stefan Mann has written ten books, among them an economics textbook, and has worked at earlier stages at the German Ministry of Agriculture and at Rostock University. His field of specialization are different aspects of multifunctionality in agriculture.

Bernadette Montanari obtained her PhD in Ethnobiology from the University of Kent, UK. Her fields of expertise and interest are: aromatic and medicinal plants, essential oil distillation and other natural resources that can potentially be developed for community enterprise, social and indigenous entrepreneurship, women and community empowerment, poverty alleviation. Her work addresses traditional ecological knowledge pathways for transmission and erosion. Further interests include ethno-medicine, ethno-pharmacology, aromatic and medicinal plants conservation, political anthropology, decentralization processes and community development issues.

Francis Mulangu is an agricultural economist at the African Center for Economic Transformation (ACET) in Accra, Ghana. He holds a Master’s degree in Economics and PhD in Agricultural, Environmental, and Development Economics. Francis Mulangu’s dissertation work used primary fieldwork data to study the impact of climate change on agriculture in the Kilimanjaro region of Tanzania.

Erich Tasser works for the European Academy. He is a trained ecologist and landscape modeler. His main research areas are alpine agriculture, sustainable development and landscape planning. Within these topics he is working on biodiversity, vegetation analysis, and ecosystem mapping in mountain regions using GIS and applied statistical methods.

Index

S. Mann (ed.), *The Future of Mountain Agriculture*, Springer Geography,
DOI: 10.1007/978-3-642-33584-6, © Springer-Verlag Berlin Heidelberg 2013

Printed by Printforce, the Netherlands